Math Achievement

Enriching Activities Based on NCTM Standards

Grade 7

by

Tracy Dankberg
and Leland Graham

Table of Contents

Introduction

Welcome to the **Math Achievement** series! Each book in this series is designed to reinforce the math skills appropriate for each grade level and to encourage high-level thinking and problem-solving skills. Enhancing students' thinking and problem-solving abilities can help them succeed in all academic areas. In addition, experiencing success in math can increase a student's confidence and self-esteem, both in and out of the classroom.

Each **Math Achievement** book offers challenging questions **based on the standards specified by the National Council of Teachers of Mathematics (NCTM)**. All five content standards (number and operations, algebra, geometry, measurement, data analysis and probability) and the process standard, problem solving, are covered in the activities.

The questions and format are similar to those found on standardized math tests. The experience students gain from answering questions in this format may help increase their test scores.

These exercises can be used to enhance the regular math curriculum, to individualize instruction, to provide extra practice for home schoolers, or to review skills between grades.

The following math skills are covered in this book:

- **problem solving**
- **rounding**
- **estimation**
- **multiplication**
- **division**
- **integers**
- **fractions**
- **decimals**
- **ratio and percent**
- **algebra**
- **geometry**
- **tables and graphs**
- **probability & statistics**

Each **Math Achievement** book contains **four pretests in standardized test format** at the beginning of each book. The pretests have been designed so that they may be used individually, as four stand-alone tests, or in groups. They may be used to identify students' needs in specific areas, or to compare students' math abilities at the beginning and end of the school year. **A scoring box is also included on each activity page.** This scoring box can be programmed to suit your specific classroom and student needs with total problems, total correct, and score.

Name ______________________________ **Pretest**

Read each problem. Circle the letter beside the correct answer.

1. Rounded to the nearest ten, 47 is closest to _____.
   A. 30 B. 50 C. 70 D. 40

2. Rounded to the nearest hundred, 289 is closest to _____.
   A. 280 B. 100 C. 300 D. 200

3. Rounded to the nearest thousand, 6,127 is closest to _____.
   A. 5,000 B. 6,000 C. 5,900 D. 4,000

4. David Hall Middle School has 1,081 students enrolled. Three hundred ninety-one students are seventh graders. Estimate to the nearest hundred how many students are not seventh graders.
   A. 400 B. 500 C. 600 D. 700

5. In its first year, the school band had 93 members. This year, it has 205 members. Use front-end estimation to tell how many new members joined this year.
   A. 110 B. 150 C. 125 D. 100

6. Estimate 485 ÷ 7 = _____. A. 60 B. 50 C. 70 D. 40

7. Estimate 3,576 ÷ 4 = _____. A. 900 B. 600 C. 700 D. 800

8. Estimate 152 x 6 = _____. A. 940 B. 900 C. 600 D. 650

9. Estimate 93 x 29 = _____. A. 3,200 B. 2,400 C. 320 D. 2,700

10. Suppose 551 Academic Team members competed in a Tournament Championship. If there are 6 members on a team, how many teams are competing? Estimate your answer.
    A. 90 B. 80 C. 95 D. 70

11. 5^4 is another way of writing _____.
    A.5 x 4 C. 4 x 5
    B.4 x 4 x 4 x 4 x 4 D. 5 x 5 x 5 x 5

12. The expression 6^3 = _____.
    A. 36 B. 18 C. 666 D. 216

13. The expression 2 x 2 x 2 x 2 can also be written as _____.
    A. 2 x 4 B. 4 x 2 C. 2^4 D. 4^2

Read each problem. Circle the letter beside the correct answer.

Total Problems: **Total Correct:** **Score:**

Name ______________________________ **Pretest**

Read each problem. Circle the letter beside the correct answer.

1. What is the value of this expression? 36 ÷ 6 x 3 – 7
A. 10 B. 11 C. 9 D. 8

2. In this expression, which operation should be done first? 25 ÷ 3 x 7 + 4
A. addition B. subtraction C. multiplication D. division

3. What sign makes this a true sentence? 5.1 ______ 5.12
A. = B. < C. >

4. Which decimal is the greatest?
A. 1.71 B. 1.60 C. 1.17

5. Which shows the correct order, from least to greatest?
A. 4.002, 4.02, 4.2 B. 1.31, 1.13, 3.13 C. 2.1, 3.46, .76

6. 2.468 rounded to the nearest tenth is ______.
A. 2.9 B. 2.5 C. 2.8 D. 3.0

7. 69.15 rounded to the nearest tenth is ______.
A. 69.1 B. 69.0 C. 69.5 D. 69.2

8. $\begin{array}{r} 3.43 \\ +\ 1.14 \\ \hline \end{array}$
A. 4.89
B. 5.61
C. 4.69
D. 4.57

9. $\begin{array}{r} 5.687 \\ -\ 3.29 \\ \hline \end{array}$
A. 3.257
B. 2.397
C. 2.982
D. 3.278

10. Terry wants to buy a concert ticket for $47.50 plus a $2.50 service charge. If he pays for it with three $20 bills, estimate the amount of change he should receive.
A. $9.00 B. $12.00 C. $20.00 D. $13.50

11. James is getting ready to go to Hawaii. He wants two pairs of sunglasses that cost $15.99 each, a scuba suit that costs $49.79, a cap that costs $6.49, and a bottle of sunscreen that costs $8.98. What is the best estimate of the cost of the items James wants to buy?
A. $82.00 B. $97.00 C. $85.00 D. $81.00

12. Which shows the fractions in order from least to greatest?
A. $\frac{5}{8}, \frac{1}{2}, \frac{1}{4}, \frac{7}{8}$ B. $\frac{1}{2}, \frac{3}{4}, \frac{7}{8}, \frac{1}{4}$ C. $\frac{1}{4}, \frac{1}{2}, \frac{3}{4}, \frac{7}{8}$ D. $\frac{1}{3}, \frac{1}{4}, \frac{1}{2}, \frac{7}{8}$

13. Which fraction has been reduced to lowest terms?
A. $\frac{6}{16}$ B. $\frac{3}{12}$ C. $\frac{2}{4}$ D. $\frac{5}{8}$

Name ________________________________

Read each problem. Circle the letter beside the correct answer.

1. $\frac{3}{5} + \frac{3}{5} =$ ______. A. $1\frac{1}{5}$ B. $\frac{8}{5}$ C. 1 D. $1\frac{2}{5}$

2. Julie's mom estimates that she will need $2\frac{7}{8}$ yards of fabric to make her prom dress and $1\frac{3}{8}$ yards to make a jacket to go with it. About how much fabric does she need?
   A. $4\frac{5}{8}$ yards B. $3\frac{7}{8}$ yards C. $4\frac{1}{4}$ yards D. $4\frac{1}{2}$ yards

3. Jay's score on the Spanish test was 93%. How would this be written as a decimal?
   A. 9.3 B. .93 C. .903 D. 9.03

4. Written as a decimal, 33% would be ______.
   A. 0.33 B. 3.03 C. 33.0 D. 0.3

5. The best estimate of 40% of 248 is ______.
   A. 60 B. 120 C. 100 D. 75

6. In a survey, researchers asked 3,000 people if they bought or rented more movies. If 20% said they bought them, about how many rented their movies?
   A. 240 B. 600 C. 6,000 D. 2,400

7. Which ratio correctly shows that 78 of 192 students in the seventh grade are girls?
   A. 78 > 192 B. 78 : 192 C. 1 out of 4 D. 1 : 2

8. When expressed as a rate, the ratio 336 miles on 14 gallons of gas would be ______.
   A. 24 miles per gallon B. 38 miles per gallon C. 34 miles per gallon

9. Which of the following pairs of ratios correctly expresses a proportion?
   A. $\frac{8}{10}, \frac{80}{100}$ B. $\frac{6}{5}, \frac{12}{15}$ C. $\frac{3}{7}, \frac{9}{21}$ D. $\frac{2}{3}, \frac{5}{6}$

10. Marsha is making trail mix for a party. She needs a can of peanuts at $4.99, one box of raisins at $1.79, and one bag of candies at $3.89. She has a coupon to save 75¢ on the candy. Sales tax is 5%. How much will the trail mix cost to make?
    A. $16.06 B. $15.66 C. $10.92 D. $10.42

11. If x = 12 and y = 28, solve for z in the following expression: z – x = y.
    A. 30 B. 26 C. 40 D. 50

12. If m = 9, solve for n in the following expression: 6m – n = 14.
    A. 27 B. 30 C. 40 D. 12

Total Problems: **Total Correct:** **Score:**

Name ______________________________ **Pretest**

Read each problem. Circle the letter beside the correct answer.

1. Beth bought a new pair of running shoes and three pairs of socks. She spent a total of $92. The running shoes cost $79. Which equation could be used to find the cost of the socks if **s** represents the cost of the socks?
   A. $79 + $92 = 3s   B. $79 + 3s = $92   C. $79 – 3s = 92

---

2. Which is an obtuse angle?

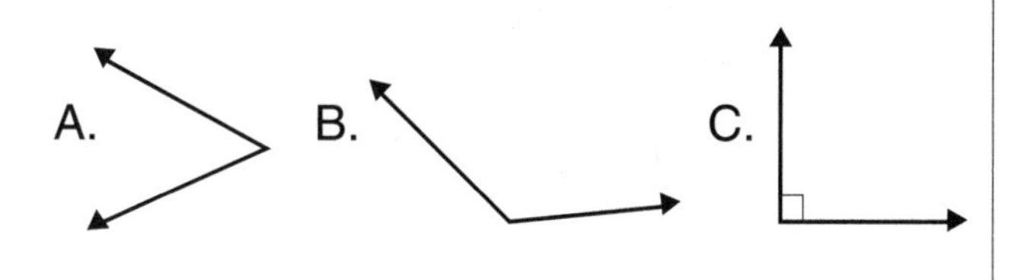

3. Which is an equilateral triangle?

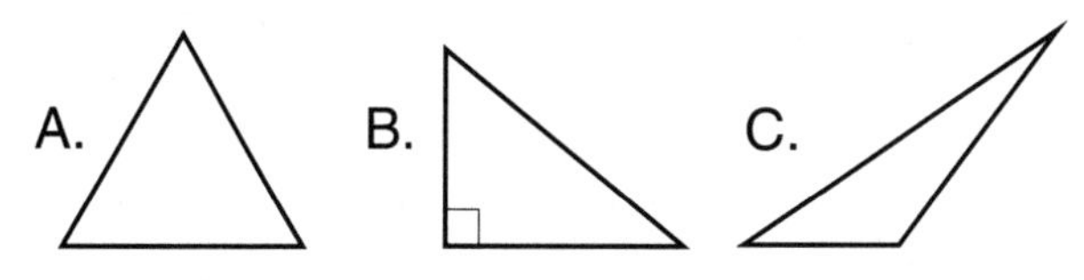

---

4. What is the area of this triangle? $A = \frac{1}{2}(b \times h)$
   A. 72 square inches
   B. 36 square inches
   C. 32 square inches
   D. 40 square inches

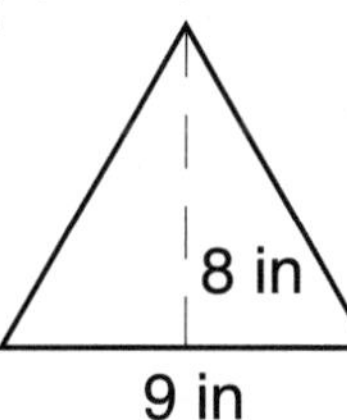

---

5. What is the area of this circle? $A = 3.14 \times r^2$
   A. 18.84 square inches
   B. 36.23 square inches
   C. 113.04 square inches
   D. 28.26 square inches

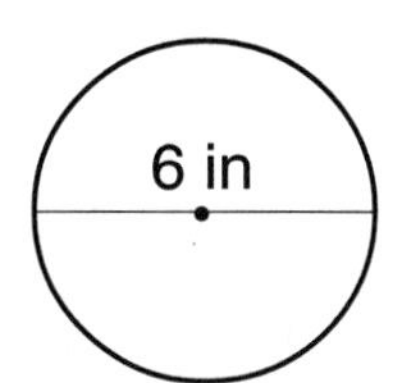

Use the stem and leaf plot to answer the following questions. The digits to the left of the line have the greater place value and are called stems. Digits to the right of the line represent digits in the ones place and are called leaves.

stem	leaf
4	566789
5	001112334555
6	012224566

6. How many times is the number 51 represented in the chart?
   A. 1   B. 2   C. 3   D. 4

7. The smallest number represented is ______.
   A. 45   B. 48   C. 40   D. 42

8. The range of numbers represented is ______.
   A. 49-60   B. 50-69   C. 45-66   D. 40-70

Total Problems:	Total Correct:	Score:

Pretest Answer Key

Name ______________________ **Pretest**

Read each problem. Circle the letter beside the correct answer.

1. Rounded to the nearest ten, 47 is closest to _____.
   A. 30 Ⓑ 50 C. 70 D. 40
2. Rounded to the nearest hundred, 289 is closest to _____.
   A. 280 B. 100 Ⓒ 300 D. 200
3. Rounded to the nearest thousand, 6,127 is closest to _____.
   A. 5,000 Ⓑ 6,000 C. 5,900 D. 4,000
4. David Hall Middle School has 1,081 students enrolled. Three hundred ninety-one students are seventh graders. Estimate to the nearest hundred how many students are not seventh graders.
   A. 400 B. 500 C. 600 Ⓓ 700
5. In its first year, the school band had 93 members. This year, it has 205 members. Use front-end estimation to tell how many new members joined this year.
   Ⓐ 110 B. 150 C. 125 D. 100
6. Estimate 485 ÷ 7 = _____. A. 60 B. 50 Ⓒ 70 D. 40
7. Estimate 3,576 ÷ 4 = _____. Ⓐ 900 B. 600 C. 700 D. 800
8. Estimate 152 x 6 = _____. A. 940 Ⓑ 900 C. 600 D. 650
9. Estimate 93 x 29 = _____. A. 3,200 B. 2,400 C. 320 Ⓓ 2,700
10. Suppose 551 Academic Team members competed in a Tournament Championship. If there are 6 members on a team, how many teams are competing? Estimate your answer.
    Ⓐ 90 B. 80 C. 95 D. 70
11. 5^4 is another way of writing _____.
    A.5 x 4 C. 4 x 5
    B.4 x 4 x 4 x 4 x 4 Ⓓ 5 x 5 x 5 x 5
12. The expression 6^3 = _____.
    A. 36 B. 18 C. 666 Ⓓ 216
13. The expression 2 x 2 x 2 x 2 can also be written as _____.
    A. 2 x 4 B. 4 x 2 Ⓒ 2^4 D. 4^2

Read each problem. Circle the letter beside the correct answer.

Total Problems: Total Correct: Score:

Name ______________________ **Pretest**

Read each problem. Circle the letter beside the correct answer.

1. What is the value of this expression? 36 ÷ 6 x 3 – 7
   A. 10 Ⓑ 11 C. 9 D. 8
2. In this expression, which operation should be done first? 25 ÷ 3 x 7 + 4
   A. addition B. subtraction C. multiplication Ⓓ division
3. What sign makes this a true sentence? 5.1 _____ 5.12
   A. = Ⓑ < C. >
4. Which decimal is the greatest?
   Ⓐ 1.71 B. 1.60 C. 1.17
5. Which shows the correct order, from least to greatest?
   Ⓐ 4.002, 4.02, 4.2 B. 1.31, 1.13, 3.13 C. 2.1, 3.46, .76
6. 2.468 rounded to the nearest tenth is _____.
   A. 2.9 Ⓑ 2.5 C. 2.8 D. 3.0
7. 69.15 rounded to the nearest tenth is _____.
   A. 69.1 B. 69.0 C. 69.5 Ⓓ 69.2
8. 3.43 + 1.14
   A. 4.89 B. 5.61 C. 4.69 Ⓓ 4.57
9. 5.687 – 3.29
   A. 3.257 Ⓑ 2.397 C. 2.982 D. 3.278
10. Terry wants to buy a concert ticket for $47.50 plus a $2.50 service charge. If he pays for it with three $20 bills, estimate the amount of change he should receive.
    Ⓐ $9.00 B. $12.00 C. $20.00 D. $13.50
11. James is getting ready to go to Hawaii. He wants two pairs of sunglasses that cost $15.99 each, a scuba suit that costs $49.79, a cap that costs $6.49, and a bottle of sunscreen that costs $8.98. What is the best estimate of the cost of the items James wants to buy?
    A. $82.00 B. $97.00 C. $85.00 Ⓓ $81.00
12. Which shows the fractions in order from least to greatest?
    A. $\frac{5}{8}, \frac{1}{2}, \frac{1}{4}, \frac{7}{8}$ B. $\frac{1}{2}, \frac{3}{4}, \frac{7}{8}, \frac{1}{4}$ Ⓒ $\frac{1}{4}, \frac{1}{2}, \frac{3}{4}, \frac{7}{8}$ D. $\frac{1}{3}, \frac{1}{4}, \frac{1}{2}, \frac{7}{8}$
13. Which fraction has been reduced to lowest terms?
    A. $\frac{6}{16}$ B. $\frac{3}{12}$ C. $\frac{2}{4}$ Ⓓ $\frac{5}{8}$

Total Problems: Total Correct: Score:

Name ______________________ **Pretest**

Read each problem. Circle the letter beside the correct answer.

1. $\frac{3}{5} + \frac{3}{5}$ = _____. Ⓐ $1\frac{1}{5}$ B. $\frac{8}{5}$ C. 1 D. $1\frac{2}{5}$
2. Julie's mom estimates that she will need $2\frac{7}{8}$ yards of fabric to make her prom dress and $1\frac{3}{8}$ yards to make a jacket to go with it. About how much fabric does she need?
   A. $4\frac{5}{8}$ yards B. $3\frac{7}{8}$ yards Ⓒ $4\frac{1}{4}$ yards D. $4\frac{1}{2}$ yards
3. Jay's score on the Spanish test was 93%. How would this be written as a decimal?
   A. 9.3 Ⓑ .93 C. .903 D. 9.03
4. Written as a decimal, 33% would be _____.
   Ⓐ 0.33 B. 3.03 C. 33.0 D. 0.3
5. The best estimate of 40% of 248 is _____.
   A. 60 B. 120 Ⓒ 100 D. 75
6. In a survey, researchers asked 3,000 people if they bought or rented more movies. If 20% said they bought them, about how many rented their movies?
   A. 240 B. 600 C. 6,000 Ⓓ 2,400
7. Which ratio correctly shows that 78 of 192 students in the seventh grade are girls?
   A. 78 > 192 Ⓑ 78 : 192 C. 1 out of 4 D. 1 : 2
8. When expressed as a rate, the ratio 336 miles on 14 gallons of gas would be _____.
   Ⓐ 24 miles per gallon B. 38 miles per gallon C. 34 miles per gallon
9. Which of the following pairs of ratios correctly expresses a proportion?
   Ⓐ $\frac{8}{10}, \frac{80}{100}$ B. $\frac{6}{5}, \frac{12}{15}$ C. $\frac{3}{7}, \frac{9}{22}$ D. $\frac{2}{3}, \frac{5}{6}$
10. Marsha is making trail mix for a party. She needs a can of peanuts at $4.99, one box of raisins at $1.79, and one bag of candies at $3.89. She has a coupon to save 75¢ on the candy. Sales tax is 5%. How much will the trail mix cost to make?
    A. $16.06 B. $15.66 C. $10.92 Ⓓ $10.42
11. If x = 12 and y = 28, solve for z in the following expression: z – x = y.
    A. 30 B. 26 Ⓒ 40 D. 50
12. If m = 9, solve for n in the following expression: 6m – n = 14.
    A. 27 B. 30 Ⓒ 40 D. 12

Total Problems: Total Correct: Score:

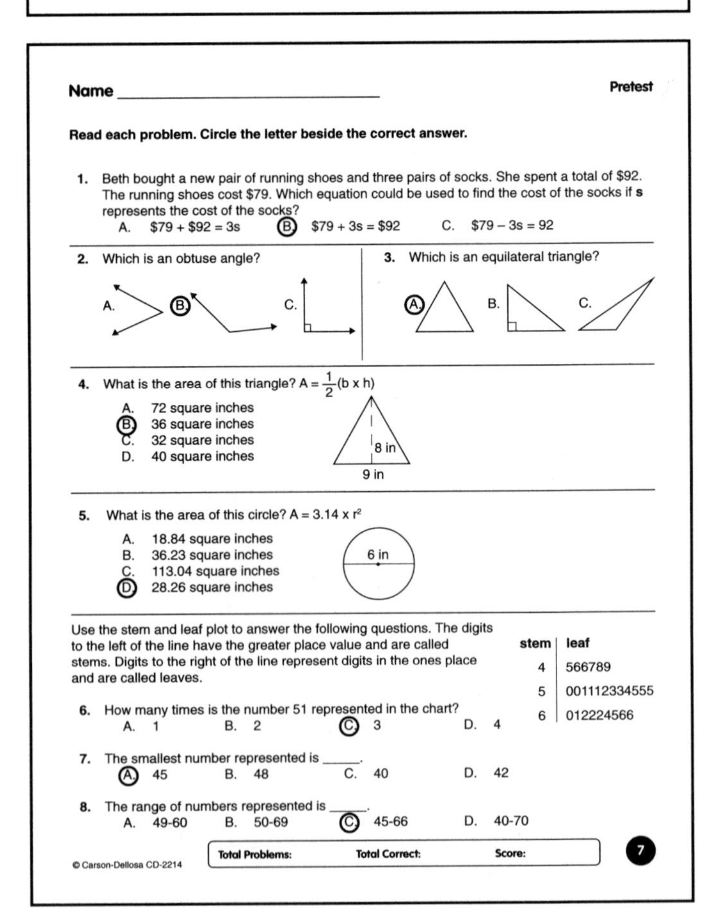

Name ______________________ **Pretest**

Read each problem. Circle the letter beside the correct answer.

1. Beth bought a new pair of running shoes and three pairs of socks. She spent a total of $92. The running shoes cost $79. Which equation could be used to find the cost of the socks if **s** represents the cost of the socks?
   A. $79 + $92 = 3s Ⓑ $79 + 3s = $92 C. $79 – 3s = 92
2. Which is an obtuse angle?
   A. Ⓑ C.
3. Which is an equilateral triangle?
   Ⓐ B. C.
4. What is the area of this triangle? $A = \frac{1}{2}(b \times h)$
   A. 72 square inches
   Ⓑ 36 square inches
   C. 32 square inches
   D. 40 square inches
   (8 in, 9 in)
5. What is the area of this circle? $A = 3.14 \times r^2$
   A. 18.84 square inches
   B. 36.23 square inches
   C. 113.04 square inches
   Ⓓ 28.26 square inches
   (6 in)

Use the stem and leaf plot to answer the following questions. The digits to the left of the line have the greater place value and are called stems. Digits to the right of the line represent digits in the ones place and are called leaves.

stem	leaf
4	566789
5	001112334555
6	012224566

6. How many times is the number 51 represented in the chart?
   A. 1 B. 2 Ⓒ 3 D. 4
7. The smallest number represented is _____.
   Ⓐ 45 B. 48 C. 40 D. 42
8. The range of numbers represented is _____.
   A. 49-60 B. 50-69 Ⓒ 45-66 D. 40-70

Total Problems: Total Correct: Score:

Name ________________________________

Addition and Subtraction
Mixed Practice

Study the box below. Find each sum or difference and write the answer in the space provided.

Rule:	**Example:**
Line up the numbers. Start with the ones place. Carry or borrow if necessary.	10,956 + 3,117 **14,073**

1. 5,746
+ 6,832

2. 6,502
+ 3,686

3. 800
− 659

4. 495
72
+ 354

5. 7,465
+ 9,567

6. 675
+ 786

7. 70,000
− 51,902

8. 4,692
745
+ 3,026

9. 2,794
+ 4,556

10. 7,000
− 425

11. 3,673
− 2,094

12. 775
783
+ 367

13. 7,400
− 6,027

14. 6,245
+ 4,414

15. 2,429
− 858

16. 23,651
56,225
+ 36,568

17. 19,000
− 3,767

18. 3,025
− 1,675

19. 3,672
+ 8,074

20. 41,908
34,821
+ 27,889

Total Problems: **Total Correct:** **Score:**

Name ________________________________

Study the examples below. Estimate each sum or difference by rounding to the highest place value both numbers have in common. Write the answer in the space provided.

Examples:

359	→ 360	941	→ 900	2,679	→ 2,700
+ 32	→ + 30	− 267	→ − 300	+ 135	→ + 100
	390		**600**		**2,800**

1. 251 + 29

2. 327 + 19

3. 72 − 58

4. 684 + 65

5. 157 − 68

6. 792 − 481

7. 319 + 495

8. 898 − 375

9. 2,746 + 3,049

10. 5,982 − 455

11. 7,112 + 448

12. 8,613 − 4,588

13. 17,659 − 5,923

14. 14,295 + 475

15. 24,761 − 2,949

16. 31,678 + 14,902

Total Problems: Total Correct: Score:

Name ______________________

Clustering and Front-End Estimation

Study the example below. Estimate each sum using clustering. Write the answer in the space provided.

Example:

```
   48
   52
   51
 + 47
  200
```

All four numbers cluster around 50.

4 x 50 = 200

Therefore, the estimated sum is **200**.

1. 73 + 70 + 67 + 68 =
2. 89 + 91 + 93 + 88 =
3. 114 + 97 + 105 + 93 =
4. 303 + 295 + 305 + 299 =
5. 79 + 83 + 78 + 81 =
6. 391 + 398 + 401 + 405 =
7. 148 + 149 + 153 + 152 =
8. 271 + 264 + 268 + 273 =
9. 191 + 190 + 188 + 187 =
10. 413 + 409 + 412 + 407 =

Study the box below. Estimate each sum or difference by using front-end estimation. Write the answer in the space provided.

Rule:

Add or subtract the front-end digits.

Adjust the number by estimating the sum or difference of the remaining digits.

Examples:

```
  519        20          873        70
+ 677      + 80        - 251      - 50
1,100       100          600        20
         + 1,100                 + 600
           1,200                   620
```

11. 788 + 459 =
12. 253 − 46 =
13. 1,750 − 549 =
14. 2,393 + 3,545 =
15. 7,913 − 4,526 =

Total Problems: Total Correct: Score:

Name ______________________________

Problem Solving with Whole Numbers

Solve each word problem. Show your work and write the answer in the space provided.

1. Of the 1,329 students at Peaks Middle School, 483 voted in favor of having school uniforms. How many students voted against it?

2. An amusement park had 39,345 people enter its gates on Saturday and 42,195 on Sunday. How many people walked through the gates in all?

3. John purchased a new car at a cost of $29,199. If he had to pay $2,045 in tax, how much did he spend for the car?

4. During the food drive, Harris's school collected 2,451 cans of assorted food. His younger sister's school collected 1,973 cans of food. How many cans of food were collected in all?

5. Gary's company had sales of $178,354 during the month of October. He had expenses of $129,459. How much profit did he make after expenses?

6. Sean's salary is $57,000 per year. Iris makes $42,191 per year. How much more money does Sean make?

7. During his high school football career, David rushed for 3,841 yards. His younger brother Joe rushed for 4,095 yards. How many more yards did Joe rush?

8. Akeem put 11,951 miles on his new car during the first year. During the second year, he drove it 8,973 miles. How many miles did Akeem have on his car after the second year?

Total Problems: **Total Correct:** **Score:**

Name ______________________________

Find the products. Write the answer in the space provided.

1. 24 x 9 =

2. 73 x 8 =

3. 42 x 6 =

4. 583 x 5 =

5. 415 x 72 =

6. 349 x 27 =

7. 4,510 x 5 =

8. 8,032 x 34 =

9. 9,316 x 48 =

10. 38 x 289 =

11. 672 x 315 =

12. 820 x 141 =

13. 6,509 x 446 =

14. 5,592 x 700 =

15. 642 x 526 =

16. 2,786 x 563 =

17. 9,017 x 483 =

18. 1,598 x 677 =

19. 2,638 x 450 =

20. 3,138 x 495 =

Total Problems: Total Correct: Score:

Name ____________________

Study the examples below. Then, divide and write the answer in the space provided.

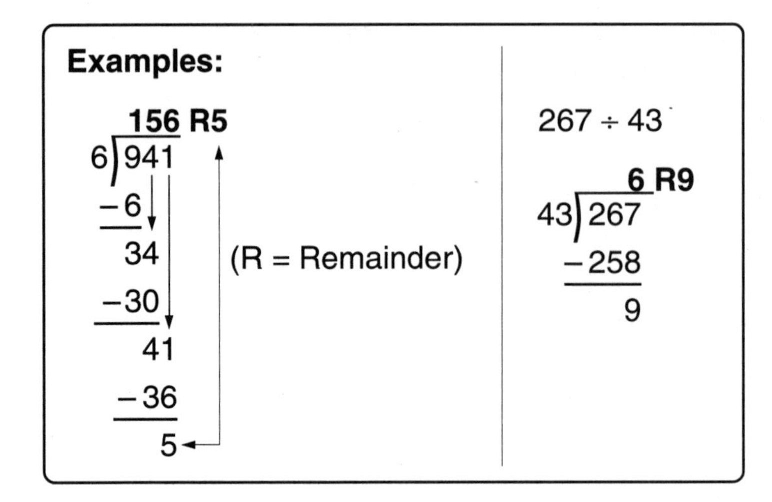

1. 5)193

2. 6)1,242

3. 7,638 ÷ 27 =

4. 17)298

5. 34)675

6. 6,590 ÷ 34 =

7. 42)1,495

8. 55)6,825

9. 54,610 ÷ 64 =

10. 62)2,657

11. 74)3,220

12. 34,293 ÷ 42 =

Total Problems: **Total Correct:** **Score:**

Name ______________________________

Study the example below. Use rounding to estimate each product. Round to the highest place value but do not round single-digit numbers. Write the answer on the line provided.

Example:

275 x 9
↓ ↓
300 x 9   3 x 9 = 27 plus 2 zeros, gives an estimate of **2,700**.

1. 28 x 7 = ______________________

2. 21 x 17 = ______________________

3. 185 x 2 = ______________________

4. 218 x 41 = ______________________

5. 1,321 x 63 = ______________________

6. 2,812 x 718 = ______________________

7. 15 x 19 = ______________________

8. 39 x 41 = ______________________

9. 231 x 7 = ______________________

10. 392 x 115 = ______________________

11. 2,499 x 83 = ______________________

12. 5,081 x 308 = ______________________

13. 43 x 8 = ______________________

14. 72 x 45 = ______________________

15. 93 x 121 = ______________________

16. 451 x 242 = ______________________

17. 25 x 749 = ______________________

18. 6,250 x 627 = ______________________

Total Problems: **Total Correct:**   **Score:**

Name ______________________________

Study the box below. Use compatible numbers to estimate each quotient. Write the answer on the line provided.

Rule:	**Example:**
To estimate with compatible numbers, find 2 numbers that are close to the originals. Then, divide without a remainder.	$2,379 \div 59$ ↓ ↓ $2,400 \div 60 = 240 \div 6 = \mathbf{40}$

1. $163 \div 4 =$ ______________

2. $712 \div 9 =$ ______________

3. $432 \div 6 =$ ______________

4. $534 \div 9 =$ ______________

5. $245 \div 8 =$ ______________

6. $1,230 \div 6 =$ ______________

7. $2,191 \div 7 =$ ______________

8. $2,819 \div 39 =$ ______________

9. $4,889 \div 71 =$ ______________

10. $3,591 \div 62 =$ ______________

11. $6,338 \div 78 =$ ______________

12. $2,412 \div 61 =$ ______________

13. $4,308 \div 72 =$ ______________

14. $2,553 \div 48 =$ ______________

Total Problems: **Total Correct:** **Score:**

Name ____________________

Study the box below. Evaluate each expression. Write the answer on the line provided.

Rule:	**Examples:**	
Order of Operations:	$5^2 + 3 \bullet 2$	$17 - 5 \bullet 3$
1. Work inside the parentheses.	$25 + 3 \bullet 2$	$17 - 15 =$ **2**
2. Compute all exponents.	$25 + 6 =$ **31**	
3. Multiply and divide from left to right.		
4. Add and subtract from left to right.		

1. $2 + 5 \bullet 2^2 =$ ____________

2. $(2 + 5) \bullet 2^2 =$ ____________

3. $(3^3 - 5) \bullet 2 =$ ____________

4. $3^3 - 5 \bullet 2 =$ ____________

5. $3 \bullet (5^2 - 10) =$ ____________

6. $3 \bullet 5^2 - 10 =$ ____________

7. $25 \div (2 + 3)^2 =$ ____________

8. $30 \div 2 + 2^2 =$ ____________

9. $(8 - 2)^2 \bullet 2 =$ ____________

10. $8 - 1 \bullet 4 + 3 =$ ____________

11. $(8 - 1) \bullet 4 + 3 =$ ____________

12. $(6^2 - 3) + 5 =$ ____________

13. $4 + (2^2 - 3) + 5 =$ ____________

14. $6^2 - 6 \div 3 =$ ____________

Total Problems: **Total Correct:** **Score:**

Name ____________________

Study the box below. Write each problem in expanded form on the line provided.

Rule:	**Examples:**	
An exponent tells the number of times a base is multiplied by itself.	$4^3 = 4 \bullet 4 \bullet 4$	Expanded form
$6 \bullet 6 \bullet 6 \bullet 6 \bullet 6 = 6^5$ (5 → **exponent**, 6 → **base**)	$2 \bullet 4 \bullet 2 \bullet 4 = 2^2 \bullet 4^2$	Exponential form
	$5^2 = 25$	Simplified

1. 10^2 ____________ **4.** 12^3 ____________ **7.** 6^2 ____________

2. 7^3 ____________ **5.** 4^6 ____________ **8.** c^3 ____________

3. 3^5 ____________ **6.** 9^7 ____________ **9.** a^4 ____________

Write each problem in exponential form on the line provided.

10. $9 \bullet 9 \bullet 9$ ____________ **13.** $6 \bullet 6 \bullet 5$ ____________ **16.** $4 \bullet 5 \bullet 4 \bullet 7$ ____________

11. $11 \bullet 11$ ____________ **14.** $2 \bullet 2 \bullet 3 \bullet 2 \bullet 3$ ____________ **17.** $y \bullet y \bullet y$ ____________

12. $13 \bullet 13 \bullet 13 \bullet 13$ ____________ **15.** $11 \bullet 12 \bullet 11$ ____________ **18.** $n \bullet n \bullet n \bullet n$ ____________

Write each problem in simplified form on the line provided.

19. 6^2 ____________ **21.** $2 \bullet 3^2$ ____________ **23.** $10^2 \bullet 5^3$ ____________

20. 4^3 ____________ **22.** $3^3 + 7^2$ ____________ **24.** $2^5 + 4^3$ ____________

Total Problems: **Total Correct:** **Score:**

Name ______________________________

Solve each word problem. Show your work and write the answer in the space provided.

1. The seventh grade basketball team is planning a banquet. There will be 142 people at the banquet. If each table holds 8 people, how many tables will be needed?

2. Gilbert saw an advertisement in the paper for golf lessons that read: "12 lessons for $35 per lesson." If Sean decides to buy all 12 lessons, how much will he spend?

3. Sarah has a part-time job after school. She earns $75 per week. If she works all but 3 weeks during the year, how much money will she earn in 1 year?

4. A trip from Atlanta, Georgia to Ft. Lauderdale, Florida is approximately 600 miles by car. If Debbie's car averages 20 miles per gallon, how many gallons of gasoline will she use for a round trip?

5. Five hundred fifty people attended a $250 per plate fund-raiser for a politician. How much money did the politician raise?

6. Ray wanted to improve his golf swing. He went out to a driving range every day for 56 days straight. How many weeks was that?

7. Natalia drives an average of 35 miles per day for work. How many miles does she drive in 1 week? (She works Monday–Friday.) How many miles will she drive in 1 year, if she takes a 2-week vacation?

8. Susan walked onto an elevator and noticed a posted sign that read: "Maximum capacity 2,000 pounds." How many people, weighing an average of 150 pounds, can fit on the elevator?

Total Problems: **Total Correct:** **Score:**

Name ______________________

Study the box below. Find each absolute value. Write the answer on the line provided.

Rule:	**Examples:**	
The absolute value of a number is its distance from 0. The following symbol is used with absolute value: $\mid \ \mid$.	$\mid {}^-9 \mid = \mathbf{9}$	${}^-9$ is 9 places from 0, so its absolute value is 9.
	$\mid 23 \mid = \mathbf{23}$	23 is 23 places from 0, so its absolute value is 23.

1. $\mid {}^-5 \mid =$ __________
2. $\mid 8 \mid =$ __________
3. $\mid {}^-10 \mid =$ __________
4. $\mid 15 \mid =$ __________
5. $\mid 31 \mid =$ __________
6. $\mid {}^-7 \mid =$ __________

Study the box below. Find each sum. Write the answer on the line provided.

Rules:	**Examples:**	
The sum of 2 positive integers is positive. The sum of 2 negative integers is negative. When 1 integer is positive and 1 integer is negative, subtract their absolute values and use the sign of the greater absolute value.	$8 + 15 = \mathbf{23}$ ${}^-4 + {}^-9 = \mathbf{{}^-13}$	${}^-9 + 7 =$ $\mid {}^-9 \mid - \mid 7 \mid =$ ↓ ↓ $9 - 7 = 2$ Give 2 a negative value. Therefore, ${}^-9 + 7 = \mathbf{{}^-2}$.

7. ${}^-9 + 15 =$ __________
8. $24 + 7 =$ __________
9. $({}^-12) + ({}^-19) =$ __________
10. $({}^-17) + ({}^-5) =$ __________
11. ${}^-20 + 15 =$ __________
12. ${}^-8 + 2 =$ __________
13. $0 + ({}^-22) =$ __________
14. ${}^-40 + 21 =$ __________
15. $17 + ({}^-8) =$ __________

Total Problems: Total Correct: Score:

Name ______________________________

Study the box below. Compare using "<," ">," or "=." Write the answer in the box provided.

Rule:

You can use a number line to compare integers. The integer that is farther to the right on the number line has the greater value.

Example:

3 [>] -3

3 is farther to the right, so **3 > -3**.

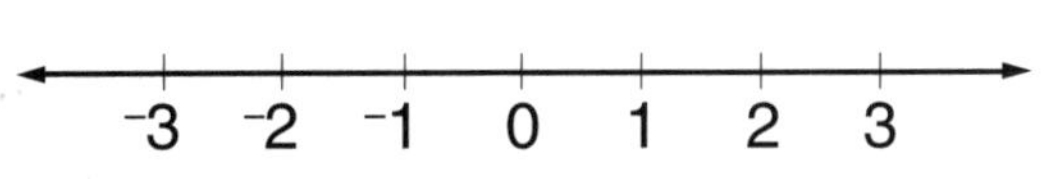

1. -8 ☐ -18
2. -2 ☐ -7
3. -7 ☐ -6
4. -50 ☐ 50
5. -68 ☐ -687
6. -13 ☐ -131
7. -8 ☐ -8
8. -73 ☐ 3
9. 2 ☐ -2
10. -54 ☐ -45
11. -27 ☐ -28
12. -15 ☐ -13

Order the sets of numbers from least to greatest. Write the answer on the line provided.

13. -3, -7, 3, 0, -8, -4

____

14. 2, -9, 3, 9, -3

____

15. 25, -25, 30, -30, -40

____

16. 3, -10, -77, -92, 42, 19

____

17. 40, -40, -10, 10, 0, 15

____

18. -125, 125, 130, -135, 140

____

Total Problems: **Total Correct:** **Score:**

Name ________________________________

Study the box below. Find each difference. Write the answer on the line provided.

Rule:	**Examples:**		
To subtract integers, add the opposite.	$10 - (^{-}7)$	$^{-}14 - 20$	$^{-}21 - (^{-}9)$
	$10 + 7 = \mathbf{17}$	$^{-}14 + {}^{-}20 = \mathbf{^{-}34}$	$^{-}21 + 9 = \mathbf{^{-}12}$

1. $5 - (^{-}16) =$ ____________________

2. $^{-}7 - 8 =$ ____________________

3. $8 - (^{-}30) =$ ____________________

4. $7 - 14 =$ ____________________

5. $45 - (^{-}20) =$ ____________________

6. $2 - 10 =$ ____________________

7. $11 - 13 =$ ____________________

8. $64 - (^{-}8) =$ ____________________

9. $^{-}13 - 15 =$ ____________________

10. $^{-}4 - (^{-}6) =$ ____________________

11. $^{-}13 - {}^{-}57 =$ ____________________

12. $5 - (^{-}55) =$ ____________________

13. $20 - (^{-}20) =$ ____________________

14. $16 - (^{-}4) =$ ____________________

15. $^{-}68 - (^{-}68) =$ ____________________

16. $^{-}40 - 25 =$ ____________________

17. $2 - 76 =$ ____________________

18. $5 - (^{-}13) =$ ____________________

19. $^{-}17 - (^{-}17) =$ ____________________

20. $32 - 100 =$ ____________________

21. $^{-}7 - (^{-}20) =$ ____________________

22. $24 - (^{-}22) =$ ____________________

Total Problems:	**Total Correct:**	**Score:**

Name ______________________________

Multiplying Integers

Study the box below. Find each product. Write the answer on the line provided.

Rule:	**Examples:**	
When the signs are the same (both positive or both negative), the answer will be positive.	$^{-}4(^{-}8) = \mathbf{32}$	$6 \times 7 = \mathbf{42}$
When the signs are different (1 positive and 1 negative), the answer will be negative.	$5(^{-}8) = \mathbf{^{-}40}$	$^{-}9(10) = \mathbf{^{-}90}$

1. $9(^{-}3) =$ __________

2. $100(12) =$ __________

3. $13(^{-}9) =$ __________

4. $^{-}7(20) =$ __________

5. $8(^{-}11) =$ __________

6. $12(^{-}5) =$ __________

7. $(^{-}9)(^{-}4) =$ __________

8. $6(30) =$ __________

9. $(^{-}50)(3) =$ __________

10. $^{-}7(^{-}14) =$ __________

11. $^{-}5(60) =$ __________

12. $4(^{-}40) =$ __________

13. $^{-}16(6) =$ __________

14. $(^{-}23)(^{-}5) =$ __________

15. $8(^{-}25) =$ __________

16. $(^{-}15)(^{-}9) =$ __________

17. $^{-}7(0) =$ __________

18. $^{-}5(2)(3) =$ __________

19. $^{-}1(^{-}2)(^{-}8) =$ __________

20. $6(^{-}2)(4) =$ __________

21. $2(7)(^{-}5) =$ __________

22. $^{-}3(4)(6) =$ __________

23. $^{-}4(^{-}8)(2) =$ __________

24. $(^{-}8)(^{-}10)(5) =$ __________

25. $(^{-}2)^2 =$ __________

26. $(^{-}2)^3 =$ __________

27. $(^{-}2)^4 =$ __________

Total Problems: **Total Correct:** **Score:**

Name ______________________________

Study the box below. Find each quotient. Write the answer on the line provided.

Rule:	**Examples:**	
When the signs are the same (both positive or both negative), the answer will be positive.	$^{-}49 \div (^{-}7) = \mathbf{7}$	$24 \div 3 = \mathbf{8}$
When the signs are different (1 positive and 1 negative), the answer will be negative.	$^{-}20 \div 4 = \mathbf{^{-}5}$	$64 \div (^{-}8) = \mathbf{^{-}8}$

1. $60 \div (^{-}15) =$ ________

2. $^{-}350 \div 35 =$ ________

3. $^{-}42 \div (^{-}6) =$ ________

4. $^{-}100 \div (^{-}10) =$ ________

5. $^{-}68 \div 4 =$ ________

6. $84 \div (^{-}12) =$ ________

7. $58 \div (^{-}2) =$ ________

8. $^{-}44 \div (^{-}22) =$ ________

9. $^{-}56 \div 4 =$ ________

10. $65 \div (^{-}1) =$ ________

11. $^{-}120 \div (^{-}3) =$ ________

12. $^{-}150 \div 25 =$ ________

13. $32 \div (^{-}16) =$ ________

14. $45 \div (^{-}9) =$ ________

15. $^{-}33 \div (^{-}3) =$ ________

16. $^{-}48 \div 12 =$ ________

17. $^{-}66 \div (^{-}11) =$ ________

18. $^{-}51 \div 3 =$ ________

19. $72 \div (^{-}6) =$ ________

20. $^{-}25 \div (^{-}25) =$ ________

21. $^{-}196 \div (^{-}49) =$ ________

22. $144 \div (^{-}12) =$ ________

23. $^{-}135 \div 15 =$ ________

24. $^{-}72 \div (^{-}24) =$ ________

25. $0 \div {}^{-}9 =$ ________

26. $^{-}420 \div (^{-}60) =$ ________

27. $^{-}85 \div 5 =$ ________

28. $^{-}132 \div (^{-}11) =$ ________

29. $96 \div (^{-}8) =$ ________

30. $^{-}140 \div (^{-}2) =$ ________

Total Problems: **Total Correct:** **Score:**

Name ______________________________

Solve each word problem. Show your work and write the answer in the space provided.

1. Raoul had a balance of $532 in his checking account. He wrote a check for $73. Write an addition sentence which represents this situation. How much money does he have left in his account?

2. During a golf tournament, Dave scored 2 under par ($^-2$) each day of the 4 days of the tournament. Express as an integer how many strokes he finished under par.

3. Find each product.

A. $(^-4)^2$

B. $(^-4)^3$

C. $(^-5)^2$

D. $(^-5)^3$

4. Find the product of any 2 negative factors, any 3 negative factors, any 4 factors, and any 5 negative factors. Look for a pattern in each product. Write a rule for multiplying more than 2 negative factors.

5. Ethel invested in a stock that was recommended to her by her broker. During the first 5 days of owning the stock, it dropped 10 points ($^-10$). What was the average change per day?

6. What is the sum of an integer and its opposite? Give examples to support your answer.

7. Write the next 2 terms in each sequence.

A. 3, 2, 1, 0, $^-1$, _____, _____

B. 5, 4, 2, $^-1$, _____, _____

C. $^-16$, $^-14$, $^-12$, $^-10$, _____, _____

8. When Charlotte went to bed, it was 9°F. When she woke up in the morning, the temperature had dropped to $^-2$°F. How many degrees did the temperature drop?

Total Problems: **Total Correct:** **Score:**

Name ______________________________

Equivalent Fractions

Study the box below. Find the missing numerator or denominator to make equivalent fractions. Write the answer in the space provided.

Rule:	**Example:**
Multiply (or divide) the numerator and denominator by the same number to make equivalent fractions.	$\frac{4}{5} = \frac{\_\_}{15}$ $\qquad \frac{4 \times 3}{5 \times 3} = \frac{\mathbf{12}}{\mathbf{15}}$

1. $\frac{3}{6} = \frac{\_\_}{12}$
2. $\frac{5}{11} = \frac{\_\_}{44}$
3. $\frac{3}{10} = \frac{18}{\_\_}$
4. $\frac{28}{32} = \frac{\_\_}{8}$
5. $\frac{7}{8} = \frac{49}{\_\_}$
6. $\frac{8}{15} = \frac{64}{\_\_}$
7. $\frac{24}{32} = \frac{\_\_}{8}$
8. $\frac{15}{60} = \frac{\_\_}{4}$
9. $\frac{10}{7} = \frac{70}{\_\_}$
10. $\frac{2}{3} = \frac{\_\_}{9} = \frac{8}{\_\_}$
11. $\frac{7}{8} = \frac{\_\_}{24}$
12. $\frac{6}{7} = \frac{30}{\_\_}$
13. $\frac{24}{30} = \frac{\_\_}{5}$
14. $\frac{17}{34} = \frac{2}{\_\_}$
15. $\frac{6}{13} = \frac{12}{\_\_}$
16. $\frac{3}{8} = \frac{\_\_}{40}$
17. $\frac{9}{12} = \frac{27}{\_\_}$
18. $\frac{9}{9} = \frac{\_\_}{25}$
19. $\frac{12}{4} = \frac{24}{\_\_} = \frac{\_\_}{40}$
20. $\frac{3}{4} = \frac{6}{\_\_} = \frac{\_\_}{32}$
21. $\frac{4}{9} = \frac{\_\_}{36}$
22. $\frac{8}{9} = \frac{\_\_}{45}$
23. $\frac{5}{12} = \frac{\_\_}{48}$
24. $\frac{27}{45} = \frac{\_\_}{5}$
25. $\frac{9}{11} = \frac{\_\_}{66}$
26. $\frac{18}{48} = \frac{3}{\_\_}$
27. $\frac{11}{15} = \frac{\_\_}{60}$
28. $\frac{4}{16} = \frac{\_\_}{4}$
29. $\frac{7}{9} = \frac{\_\_}{54}$
30. $\frac{9}{10} = \frac{27}{\_\_} = \frac{\_\_}{40}$

Total Problems:	Total Correct:	Score:

Name ______________________________

Study the box below. Compare using "<," ">," or "=." Write the answer in the box provided.

Rule:	**Example:**
To compare 2 fractions, find their cross products and compare.	$\frac{7}{8} \square \frac{8}{9}$ $7 \times 9 = 63$ $\frac{7}{8} \square \frac{8}{9}$ $8 \times 8 = 64$ $63 < 64$, so $\frac{7}{8} \boxed{<} \frac{8}{9}$

1. $\frac{3}{5} \square \frac{7}{9}$
2. $\frac{1}{3} \square \frac{2}{3}$
3. $\frac{5}{9} \square \frac{6}{10}$
4. $\frac{4}{5} \square \frac{8}{10}$
5. $\frac{8}{15} \square \frac{11}{12}$
6. $\frac{4}{7} \square \frac{8}{11}$
7. $\frac{1}{5} \square \frac{2}{7}$
8. $\frac{7}{9} \square \frac{9}{11}$
9. $\frac{5}{6} \square \frac{7}{10}$
10. $\frac{9}{12} \square \frac{7}{15}$
11. $\frac{5}{12} \square \frac{9}{16}$
12. $\frac{8}{16} \square \frac{4}{8}$
13. $\frac{2}{5} \square \frac{2}{3}$
14. $\frac{13}{16} \square \frac{17}{20}$
15. $\frac{4}{25} \square \frac{2}{5}$

Order from least to greatest. When comparing more than 2 fractions, it may be helpful to compare them 2 at a time. Write the answer on the line provided.

16. $\frac{1}{9}, \frac{1}{7}, \frac{1}{8}$ ______________________
17. $\frac{3}{4}, \frac{4}{5}, \frac{9}{10}, \frac{6}{7}$ ______________________
18. $\frac{5}{7}, \frac{5}{11}, \frac{5}{9}, \frac{5}{8}$ ______________________
19. $\frac{2}{3}, \frac{2}{7}, \frac{3}{5}$ ______________________
20. $\frac{7}{8}, \frac{13}{16}, \frac{9}{15}, \frac{11}{12}$ ______________________
21. $\frac{7}{2}, \frac{15}{4}, \frac{11}{3}, \frac{17}{6}$ ______________________

Total Problems: Total Correct: Score:

Name ______________________

Study the box below. Write each improper fraction as a mixed number or whole number. Write the answer on the line provided.

Rule:	**Example:**
1. Divide the numerator by the denominator. 2. If there is a remainder, put it in fraction form over the divisor. 3. Reduce fraction to lowest terms.	$\frac{31}{7}$ $\quad 7\overline{)31}$ = 4 R3; -28, 3 $\quad \mathbf{4\frac{3}{7}}$

1. $\frac{8}{7}$ ______

2. $\frac{38}{6}$ ______

3. $\frac{71}{9}$ ______

4. $\frac{36}{9}$ ______

5. $\frac{22}{3}$ ______

6. $\frac{62}{9}$ ______

7. $\frac{29}{9}$ ______

8. $\frac{56}{9}$ ______

Study the box below. On the line provided, write each mixed number or whole number as an improper fraction.

Rule:	**Example:**
1. Multiply the whole number by the denominator. 2. Add the numerator. 3. Place that number over the denominator.	$7\frac{3}{4}$ $\quad 7 \times 4 + 3 = \mathbf{\frac{31}{4}}$

9. $3\frac{2}{3}$ ______

10. $4\frac{9}{10}$ ______

11. 12 ______

12. $6\frac{5}{12}$ ______

13. $5\frac{7}{10}$ ______

14. 6 ______

15. $7\frac{2}{3}$ ______

16. $8\frac{3}{5}$ ______

Total Problems: **Total Correct:** **Score:**

Name ________________________________

Study the examples below. Round each fraction to 0, $\frac{1}{2}$, or 1. Write the answer on the line provided.

Examples:		
$\frac{1}{8} \rightarrow 0$	$\frac{3}{5} \rightarrow \frac{1}{2}$	$\frac{11}{12} \rightarrow 1$
The numerator is much smaller than the denominator.	The numerator is about half of the denominator.	The numerator and denominator are close in value.

1. $\frac{3}{4}$ ______

2. $\frac{1}{5}$ ______

3. $\frac{5}{8}$ ______

4. $\frac{2}{9}$ ______

5. $\frac{3}{10}$ ______

6. $\frac{7}{12}$ ______

7. $\frac{5}{6}$ ______

8. $\frac{1}{4}$ ______

9. $\frac{1}{10}$ ______

10. $\frac{7}{8}$ ______

11. $\frac{1}{7}$ ______

12. $\frac{4}{9}$ ______

Study the box below. Estimate each sum or difference. Write the answer on the line provided.

Rule:	**Example:**
Round each fraction to 0, $\frac{1}{2}$, or 1. Then, add.	$\frac{3}{8} + \frac{5}{6} =$ $\downarrow \quad \downarrow$ $\frac{1}{2} + 1 = 1\frac{1}{2}$

13. $\frac{4}{5} - \frac{1}{2} =$ ______

14. $\frac{3}{4} + \frac{1}{3} =$ ______

15. $\frac{1}{2} + \frac{7}{8} =$ ______

16. $\frac{7}{8} - \frac{1}{3} =$ ______

17. $\frac{3}{8} - \frac{1}{10} =$ ______

18. $\frac{5}{8} - \frac{1}{12} =$ ______

Total Problems: **Total Correct:** **Score:**

Name ______________________________

Study the box below. Find each sum or difference. Reduce the answer to lowest terms. Write the answer in the space provided.

Rule:

1. Change any mixed numbers to improper fractions.
2. Find the Least Common Denominator (LCD) and rewrite the fractions.
3. Add or subtract.
4. Reduce if necessary.

Examples:

$$5\frac{5}{6} = \frac{35}{6} = \frac{35}{6}$$

$$-3\frac{2}{3} = \frac{11}{3} \times \frac{2}{2} = -\frac{22}{6}$$

$$\frac{13}{6} = \mathbf{2\frac{1}{6}}$$

1. $4\frac{2}{9} + 5\frac{5}{9}$

2. $4\frac{5}{12} + 3\frac{8}{9}$

3. $5\frac{1}{5} + 5\frac{1}{3} + 3\frac{1}{15} =$

4. $7\frac{3}{8} + 5\frac{2}{5}$

5. $5\frac{1}{6} - 3\frac{2}{3}$

6. $12 - 5\frac{4}{9} =$

7. $12\frac{9}{10} - 10\frac{1}{5}$

8. $4\frac{3}{10} + 2\frac{1}{12}$

9. $5\frac{1}{4} + 6\frac{3}{10} + 3\frac{5}{8} =$

Total Problems:	Total Correct:	Score:

Name ______________________________

Study the box below. Find each product and reduce to lowest terms. Write the answer in the space provided.

Rule:	**Example:**
1. Change each mixed number to an improper fraction. 2. Multiply the numerators. 3. Multiply the denominators. 4. Reduce if possible.	$4\frac{2}{3} \times 5 =$ $\frac{14}{3} \times \frac{5}{1} = \frac{70}{3} = \mathbf{23\frac{1}{3}}$

1. $\frac{2}{5} \times \frac{3}{4} =$

2. $3\frac{2}{5} \times 4\frac{2}{3} =$

3. $1\frac{2}{3} \times 3\frac{3}{4} =$

4. $4\frac{3}{8} \times 2\frac{1}{3} =$

5. $6\frac{4}{9} \times 2\frac{2}{3} =$

6. $4\frac{1}{6} \times 3\frac{2}{3} =$

7. $5\frac{2}{5} \times 4\frac{1}{3} =$

8. $5 \times 3\frac{2}{5} =$

9. $5\frac{3}{4} \times 2\frac{1}{2} =$

10. $1\frac{4}{5} \times 2\frac{3}{10} =$

11. $5\frac{3}{8} \times 4\frac{1}{3} =$

12. $3\frac{1}{5} \times 5\frac{5}{8} =$

13. $8\frac{1}{4} \times 1\frac{1}{6} =$

14. $9 \times 4\frac{1}{4} =$

15. $3\frac{1}{4} \times 4 =$

Total Problems: **Total Correct:** **Score:**

Name ______________________________

Study the box below. Find each quotient and reduce to lowest terms. Write the answer in the space provided.

Rule:

1. Write the mixed numbers (or whole numbers) as improper fractions.
2. To divide fractions, flip the second fraction and change the division sign to multiplication.
3. Multiply, then reduce.

Remember: A whole number can be written as a fraction by placing it over 1.

Example:

$9 \div 5\frac{2}{3}$

$\frac{9}{1} \div \frac{17}{3}$

$\frac{9}{1} \times \frac{3}{17} = \frac{27}{17} = \mathbf{1\frac{27}{17}}$

1. $3\frac{1}{6} \div \frac{2}{3} =$
2. $3\frac{1}{2} \div 4\frac{3}{8} =$
3. $6\frac{1}{8} \div 2\frac{1}{4} =$
4. $7\frac{3}{10} \div 5\frac{2}{5} =$
5. $12 \div 5\frac{5}{6} =$
6. $7\frac{2}{9} \div \frac{5}{9} =$
7. $9\frac{1}{3} \div 2\frac{7}{9} =$
8. $9\frac{6}{7} \div 2\frac{2}{7} =$
9. $\frac{6}{11} \div 3\frac{4}{11} =$
10. $7\frac{8}{9} \div 3\frac{2}{3} =$
11. $21 \div 5\frac{1}{2} =$
12. $18 \div 9\frac{1}{4} =$
13. $4\frac{4}{5} \div \frac{3}{25} =$
14. $2\frac{7}{9} \div 11\frac{2}{3} =$
15. $6\frac{8}{11} \div \frac{1}{11} =$

Total Problems:	Total Correct:	Score:

Name ______________________

Solve each word problem. Show your work and write the answer in the space provided.

1. Steven works for a land developer. The developer has purchased 12 acres of land on which to build houses. If each house is to lie on a $\frac{3}{4}$-acre lot, how many houses will the developer be able to build?

2. Maria is preparing cookies for her class. Her recipe calls for $1\frac{1}{2}$ cups of flour and $\frac{2}{3}$ cup of sugar per 24 cookies. If she needs to make 8 dozen cookies, how many cups of flour will she need? How much sugar?

3. Dave spent $6\frac{1}{2}$ hours at the park on Saturday and $7\frac{3}{4}$ hours on Sunday. How many hours did he spend in all?

4. Justin needs to take a taxi to the airport. It will cost $1.75 for the first mile and $0.50 for each additional $\frac{1}{4}$ of a mile. If he is 15 miles from the airport, how much will his taxi fare cost?

5. A rectangle has a length of $1\frac{1}{2}$ centimeters and a width of $\frac{3}{4}$ centimeters. Find its area and perimeter. A = l x w, P = 2(l) + 2(w)

6. Selma needs $1\frac{1}{4}$ cups of corn syrup for a pecan pie. The bottle holds $2\frac{1}{2}$ cups. How much corn syrup will be left over?

Total Problems: **Total Correct:** **Score:**

Name ______________________________

Study the box below. Round to the indicated place-value position. Write the answer on the line provided.

Rule:	**Examples:**	
Look to the right of the indicated number. If that number is 5 or greater, round up the indicated number. If it's not, keep the number the same. Remember to drop all numbers past the rounded 1.	Tenth 34.513 → **34.5** (Drop the 1 and 3.)	Hundredth **7.915** → **7.92** (Drop the 5.)

Round to the nearest tenth.

1. 4.285 ____________

2. 38.44 ____________

3. 9.775 ____________

4. 41.36 ____________

5. 14.386 ____________

6. 45.57 ____________

7. 0.568 ____________

8. 49.814 ____________

9. 8.446 ____________

Round to the nearest hundredth.

10. 5.9998 ____________

11. 15.5458 ____________

12. 4.562 ____________

13. 3.114 ____________

14. 59.054 ____________

15. 72.986 ____________

16. 78.652 ____________

17. 27.976 ____________

18. 5.2945 ____________

Round to the nearest thousandth.

19. 52.0782 ____________

20. 82.9264 ____________

21. 80.3645 ____________

22. 3.6507 ____________

23. 2.0298 ____________

24. 20.9548 ____________

25. 63.0016 ____________

26. 11.8992 ____________

27. 37.7899 ____________

Total Problems:	Total Correct:	Score:

Name ______________________________

Study the example below. Compare using “<,” “>,” or “=.” Write the answer in the box provided.

Rule:	**Example:**
1. Line up the decimal points. 2. Compare digits from left to right in their corresponding place values. Add zeros if necessary to give each number the same number of decimal places.	0.423 [<] 0.432 0.423 0.432 $2 < 3$ so $0.423 < 0.432$

1. 9.09 ☐ 9
2. 0.78 ☐ 0.87
3. 0.001 ☐ 0.010
4. 2.582 ☐ 2.5820
5. 243.23 ☐ 243.32
6. 5.687 ☐ 56.87
7. 0.82 ☐ 0.802
8. 5.91 ☐ 5.910
9. 6.83 ☐ 6.38
10. 13.126 ☐ 13.216
11. 4.04 ☐ 4.004
12. 8.184 ☐ 8.144

Order the sets of numbers from least to greatest. Write the answer on the line provided.

13. 6.13, 6.013, 6.31 ____________________
14. 0.23, 2.03, 20.3, 0.023 ____________________
15. 28.21, 28.201, 28.021 ____________________
16. 0.415, 0.154, 0.451 ____________________
17. 8.52, 85.2, 8.05, 8.25 ____________________
18. 0.3, 3.3, 0.303, 0.03 ____________________

Total Problems: **Total Correct:** **Score:**

Name ______________________________

Study the box below. Find each sum. Write the answer in the space provided.

Rule:	**Example:**
1. Line up the decimal points.	32.5 + 39.03
2. Add zeros if necessary.	32.50
3. Add as with whole numbers.	+ 39.03
4. Carry (trade) if necessary.	**71.53**
5. Bring down the decimal point into the answer.	

1. 0.71 + 1.62

2. 23.4267 + 34.5

3. 431.51 + 273.68

4. 6.596 + 0.3 =

5. 23.73 + 62.83

6. 61.5 + 29.02

7. 645.9 + 481.765

8. 7.813 + 0.97 =

9. 363.4 + 340.8

10. 6.001 + 3.572

11. 24.093 + 7.819

12. 14 + 0.09 =

13. 815.15 + 105.07

14. 43.89 + 10.68

15. 12.708 + 43.91

16. 0.0059 + 3.2 =

Total Problems: **Total Correct:** **Score:**

Name ______________________________

Study the box below. Find each difference. Write the answer in the space provided.

Rule:	**Example:**
1. Line up the decimal points.	9.05 – 6.347
2. Add zeros if necessary.	9.050
3. Subtract as with whole numbers.	– 6.347
4. Borrow (trade) if necessary.	**2.703**
5. Bring down the decimal point into the answer.	

1. 3.08 – 0.81

2. 28.026 – 6.7

3. 0.01 – 0.001

4. 0.57 – 0.08 =

5. 45.74 – 23.88

6. 62.3 – 17.06

7. 44.74 – 39.85

8. 16.52 – 0.49 =

9. 376.3 – 242.6

10. 4.005 – 2.606

11. 0.34 – 0.108

12. 16.5 – 15 =

13. 813.1 – 66.56

14. 64.8 – 9.89

15. 5.05 – 3.768

16. 20 – 8.745 =

Total Problems: **Total Correct:** **Score:**

Name ________________________________

Estimating Sums and Differences with Decimals

Study the box below. Estimate each sum or difference using one of the strategies presented in the box. Write the strategy used and the answer in the space provided.

Examples:

Rounding

$$\begin{array}{r} 23.62 \\ -\ 6.71 \\ \hline \end{array} \longrightarrow \begin{array}{r} 24 \\ -\ 7 \\ \hline \mathbf{17} \end{array}$$

Clustering

4.1 + 3.8 + 3.9 =

4 + 4 + 4 =

4 x 3 = **12**

1. $\begin{array}{r} 63.45 \\ -\ 9.67 \\ \hline \end{array}$

2. $\begin{array}{r} 18.72 \\ +\ 15.95 \\ \hline \end{array}$

3. $\begin{array}{r} 30.09 \\ -\ 8.75 \\ \hline \end{array}$

4. $\begin{array}{r} 3.4 \\ 2.8 \\ +\ 3.3 \\ \hline \end{array}$

5. $\begin{array}{r} 27.38 \\ -\ 6.51 \\ \hline \end{array}$

6. $\begin{array}{r} 46.32 \\ +\ 59.87 \\ \hline \end{array}$

7. $\begin{array}{r} 56.25 \\ +\ 78.91 \\ \hline \end{array}$

8. $\begin{array}{r} 12.65 \\ 12.49 \\ +\ 12.58 \\ \hline \end{array}$

9. $\begin{array}{r} 78.253 \\ -\ 8.415 \\ \hline \end{array}$

10. $\begin{array}{r} 40.63 \\ +\ 59.79 \\ \hline \end{array}$

11. $\begin{array}{r} 185.34 \\ +\ 244.98 \\ \hline \end{array}$

12. $\begin{array}{r} 29.82 \\ 31.04 \\ +\ 28.88 \\ \hline \end{array}$

13. $\begin{array}{r} 270.31 \\ -\ 148.85 \\ \hline \end{array}$

14. $\begin{array}{r} 261.32 \\ +\ 158.67 \\ \hline \end{array}$

15. $\begin{array}{r} 67.15 \\ -\ 27.77 \\ \hline \end{array}$

16. $\begin{array}{r} 91.9 \\ 89.2 \\ 90.3 \\ +\ 88.9 \\ \hline \end{array}$

Total Problems: **Total Correct:** **Score:**

Name ______________________________

Study the box below. Find each product and write the answer in the space provided.

Rule:	**Example:**
Multiply as you would whole numbers. The number of decimal places in the product is the sum of the decimal places in the factors. When the problem is presented horizontally, line up the numbers on the right. Do not line up the decimal points.	Factor → .23 2 decimal places Factor → x .5 1 decimal place Product → **.115** 3 decimal places

1. 1.4 x 9

2. 2.6 x 1.2

3. 20.3 x 0.06 =

4. 48.9 x 0.95 =

5. 4.52 x 0.9

6. 3.57 x 2.4

7. 1.002 x 3.5 =

8. 17.8 x 6.518 =

9. 5.96 x 3.8

10. 6.3 x .29

11. 5.05 x 0.02 =

12. 9.6 x 35.56 =

13. 8.5 x .06

14. 16.4 x .75

15. 0.109 x 53.9 =

16. 5.06 x 0.892 =

Total Problems: **Total Correct:** **Score:**

Name ______________________________

Study the box below. Find each quotient and write the answer in the space provided.

Rule:

1. Change the divisor to a whole number by moving the decimal point to the right.
2. Move the decimal point in the dividend the same number of spaces. Add zeros if necessary.
3. Divide in the same way as with whole numbers.
4. Remember to bring up the decimal point into the quotient.

Example:

$0.7 \div 0.02$

$0.02\overline{)\,.7}$

$2\overline{)\,.70} \rightarrow 2\overline{)\,70}$ = **35**

$$\begin{array}{r} 35 \\ 2\overline{)\,70} \\ -6 \\ \hline 10 \\ -10 \\ \hline 0 \end{array}$$

1. $.4\overline{)3.6}$
2. $6.7\overline{)35.51}$
3. $0.3\overline{)15.72}$
4. $11.82 \div 0.6 =$
5. $0.7\overline{)1.05}$
6. $2.8\overline{)12.88}$
7. $4.2\overline{)2.604}$
8. $7.4 \div 0.037 =$
9. $1.1\overline{)5.5}$
10. $0.5\overline{)9.55}$
11. $1.2\overline{)104.4}$
12. $55.02 \div 4.2 =$
13. $0.46\overline{)41.4}$
14. $0.27\overline{)4.86}$
15. $0.16\overline{)0.8}$
16. $4.95 \div 0.09 =$

Total Problems: Total Correct: Score:

Name ____________________

Study the examples below. Estimate each product or quotient using the strategy indicated below. When rounding, round to the highest place value. Write the answer on the line provided.

Examples:

Estimate by rounding.

61.78 x 3.21

↓ ↓

60 x 3 = **180**

Estimate using compatible numbers.

159.63 ÷ 3.91

↓ ↓

160 ÷ 4 = **40**

Estimate by rounding.

1. 6.92 x 8.45 = ____________

2. 47.4 x 22.1 = ____________

3. 29.36 x 4.08 = ____________

4. 51.04 x 2.98 = ____________

5. 5.832 x 9.5 = ____________

6. 68.9 x 10.07 = ____________

7. 0.002 x 0.03 = ____________

8. 99.9 x 9.9 = ____________

9. 11.246 x 6.8 = ____________

10. 6.75 x 19.41 = ____________

11. 42.9 x 21.87 = ____________

12. 120.28 x 1.98 = ____________

Estimate using compatible numbers.

13. 321.9 ÷ 8.2 = ____________

14. 299.3 ÷ 4.95 = ____________

15. 181.08 ÷ 30.9 = ____________

16. 1,241.5 ÷ 42.3 = ____________

17. 1,505.3 ÷ 30.4 = ____________

18. 243.6 ÷ 4.29 = ____________

19. 543.01 ÷ 9.2 = ____________

20. 636.7 ÷ 81 = ____________

21. 3,691.91 ÷ 6.2 = ____________

22. 4,495.6 ÷ 51 = ____________

23. 468.7 ÷ 6.2 = ____________

24. 492.51 ÷ 71.1 = ____________

Total Problems: **Total Correct:** **Score:**

Name ______________________________

Study the box below. Write each number in scientific notation on the line provided.

Rule:	**Examples:**
Scientific notation expresses numbers using powers of 10. 1. Move the decimal point to change the number (n) so that $1 \leq n < 10$. 2. Count how many places you moved the decimal point. Place that number as the power of 10 (positive if it moved to the left, negative if it moved to the right).	6,210,000 = $\mathbf{6.21 \times 10^{6}}$ 0.000057 = $\mathbf{5.7 \times 10^{-5}}$

1. 845,000 ______________
2. 58,000,000 ______________
3. 0.0000063 ______________
4. 0.00089 ______________
5. 16,000 ______________
6. 800,000 ______________
7. 4,000,000 ______________
8. 0.00026 ______________
9. 6,340,000 ______________
10. 27,000,000,000 ______________

Study the box below. Write each number in standard form on the line provided.

Rule:	**Examples:**
Look at the exponent on the 10. If it is positive, move the decimal point that many places to the right. If it is negative, move the decimal point that many places to the left.	3.92×10^{5} = **392,000** 4.4×10^{-5} = **0.000044**

11. 6.01×10^{7} ______________
12. 2.8×10^{10} ______________
13. 4.33×10^{6} ______________
14. 9.3×10^{-8} ______________
15. 7.603×10^{2} ______________
16. 2.7×10^{-4} ______________
17. 4.5×10^{-1} ______________
18. 9.08×10^{-5} ______________

Total Problems:	Total Correct:	Score:

Name ______________________________

Solve each word problem. Show your work and write the answer in the space provided.

1. Janice, a piano teacher, charges $21.95 per lesson. If Sena has $200 to spend on lessons, how many will she be able to purchase?

2. Mr. Wilson had to pay $5.75 for each member of his family to ride the train around the mountain. His total charge was $34.50. How many people did Mr. Wilson pay for?

3. A magazine costs $2.79 at a newsstand. Terry orders a magazine subscription and pays $1.19 per issue. How much money is Terry saving per issue?

4. Tia ordered a hamburger for $1.99, large fries for $1.29, and a medium drink for $.89. How much did she spend?

5. Ivan earned $15.75 each week for mowing the lawn. If he spends $5.00 each week on ice cream and saves the rest, about how much will he have saved over a 12-week period?

6. Brittany's class had a car wash to raise money for a trip. The students charged $5.25 per car and $7.50 per van. If they washed 26 cars and 18 vans, how much money did they make?

7. Mimi is saving money to buy a car. She has $3,459.81 in her account. If she can raise $5,000, her parents will match that amount. How much more money does she need to raise?

8. The following numbers are in order from least to greatest: 275, 307, 574, 889. Without changing the order of those numbers or adding any zeros, place decimal points in the numbers so that they are in order from greatest to least.

Total Problems: **Total Correct:** **Score:**

Name ______________________________

Study the box below. Express each ratio as a fraction in simplest form. Write the answer on the line provided.

Rule:	**Example:**
A ratio is a comparison of two numbers. It is often written as a fraction in simplest form.	8 to 10 $\frac{8}{10} = \mathbf{\frac{4}{5}}$

1. 21 to 45 ______________________

2. 6 to 72 ______________________

3. 2 : 24 ______________________

4. 30 to 15 ______________________

5. 150 to 15 ______________________

6. 5 out of 65 ______________________

7. 64 : 18 ______________________

8. 6 out of 39 ______________________

Study the box below. Express each rate as a unit rate. Write the answer on the line provided.

Rules:	**Example:**
Rate: a ratio of 2 measurements with different units	\$0.80 per 20 forks
Unit rate: a rate in which the denominator is 1	\$0.80 ÷ 20 = **\$0.04 per 1 fork**

9. \$9.00 for 4 tapes ________________

10. \$150 for 2 days ________________

11. 330 miles in 5 hours ________________

12. 3,000 miles in 4 days ________________

13. 16 people in 4 cars ________________

14. \$30 for 2 shirts ________________

15. 4,500 tickets in 2 days ________________

16. 12 pounds in 6 weeks ________________

Total Problems: **Total Correct:** **Score:**

Name ______________________

Study the box below. Solve each proportion. Write the answer on the line provided.

Rule:	**Example:**
A proportion is 2 equal ratios.	$\frac{9}{15} = \frac{x}{10}$ $15x = 90$
To solve, find the cross products.	$x = 90 \div 15$
Then, divide to solve for the variable.	$\mathbf{x = 6}$

1. $\frac{3}{n} = \frac{5}{15}$ ______
2. $\frac{3}{5} = \frac{n}{25}$ ______
3. $\frac{3}{20} = \frac{n}{50}$ ______
4. $\frac{45}{20} = \frac{y}{4}$ ______
5. $\frac{6}{x} = \frac{42}{63}$ ______
6. $\frac{8}{5} = \frac{6}{x}$ ______
7. $\frac{4}{5} = \frac{d}{6}$ ______
8. $\frac{17}{8.5} = \frac{n}{0.01}$ ______
9. $\frac{n}{8} = \frac{12}{16}$ ______
10. $\frac{13}{4} = \frac{52}{y}$ ______
11. $\frac{12}{w} = \frac{8}{14}$ ______
12. $\frac{60}{75} = \frac{r}{5}$ ______
13. $\frac{x}{2.4} = \frac{4}{0.6}$ ______
14. $\frac{18}{m} = \frac{300}{24}$ ______
15. $\frac{2}{m} = \frac{5}{8.75}$ ______
16. $\frac{0.24}{y} = \frac{3}{9.6}$ ______
17. $\frac{8}{35} = \frac{n}{350}$ ______
18. $\frac{x}{4} = \frac{18}{30}$ ______
19. $\frac{x}{40} = \frac{15}{24}$ ______
20. $\frac{5}{1.6} = \frac{m}{4.8}$ ______
21. $\frac{1}{2} = \frac{x}{4.2}$ ______
22. $\frac{p}{5} = \frac{450}{4.5}$ ______
23. $\frac{7}{4} = \frac{n}{0.8}$ ______
24. $\frac{0.1}{8.2} = \frac{1.8}{x}$ ______

Total Problems: **Total Correct:**   **Score:**

Name ______________________________

Study the box below. Solve each problem by using a proportion. Round the answer to the nearest tenth. Write the answer on the line provided.

Rule:

Percent Proportion: $\frac{\text{Part}}{\text{Whole}} = \frac{\%}{100}$

Identify the part, whole, and/or percent.

Plug the numbers into the proportion and solve for the variable (part, whole, or percent).

Example:

Find 28% of 62.

$\frac{x}{62} = \frac{28}{100}$

$100x = 1{,}736$
$x = 17.4$

28% of 62 is **17.4**.

1. Find 80% of 45. ______________

2. 10 is 50% of what number? ______________

3. 12 is 25% of what number? ______________

4. 98 is what percent of 196? ______________

5. Find 70% of $80\frac{1}{2}$. ______________

6. 92 is what percent of 140? ______________

7. What number is 20% of 90? ______________

8. 16 is what % of 52? ______________

9. Find 75% of 50. ______________

10. What number is 60% of 40? ______________

11. 20% of 35 is what number? ______________

12. 1% of 18 is what number? ______________

Total Problems: **Total Correct:** **Score:**

Name ______________________

Study the box below. Find the percent of increase or decrease. Round to the nearest whole percent. Write the answer on the line provided.

Rule:

1. Subtract to find the amount of increase or decrease.
2. Write a proportion: $\frac{\text{change}}{\text{original}} = \frac{x\%}{100}$
3. Cross multiply to find the percent of change.

Example:

old: 29 (original)
new: 32
change: 32 – 29 = 3

$\frac{3}{29} = \frac{x}{100}$ 300 = 29x
x = 10%

1. old: \$11 new: \$9 ________
2. old: 14 new: 20 ________
3. old: 50 new: 40 ________
4. old: 20 new: 30 ________
5. old: \$80 new: \$40 ________
6. old: \$4.50 new: \$5.90 ________
7. old: \$278 new: \$350 ________
8. old: 70 new: 35 ________
9. old: 500 new: 300 ________
10. old: \$179 new: \$211 ________
11. old: 175 new: 149 ________
12. old: \$550 new: \$450 ________
13. old: 91 new: 89 ________
14. old: 0.67 new: 1.92 ________
15. old: 8.2 new: 10 ________
16. old: 72 new: 60 ________
17. old: \$126 new: \$150 ________
18. old: 0.95 new: 1.5 ________

Total Problems: **Total Correct:** **Score:**

Name ____________________

Study the examples below. Find the discount or sales tax to the nearest cent. Write the answer on the line provided.

Examples:	
\$90 dress 25% discount	\$14 music CD 7% sales tax
Think: What is 25% of \$90?	Think: What is 7% of \$14?
.25 x 90 = 22.5	.07 x 14 = .98
\$22.50 is the discount.	The sales tax is **\$0.98**.

1. \$39 theater ticket
   50% discount ____________

2. \$72 leather jacket
   6% sales tax ____________

3. \$39.00 toy
   7% sales tax ____________

4. \$89.00 calculator
   6.5% sales tax ____________

5. \$1,200 computer
   15% discount ____________

6. \$299 golf clubs
   5.5% sales tax ____________

7. \$59 jeans
   10% off ____________

8. \$22.00 sweatshirt
   30% discount ____________

9. \$17.50 alarm clock
   25% off ____________

10. \$9.00 videotape
    $\frac{1}{3}$ off ____________

11. \$245 ski outfit
    25% off ____________

12. \$125 in-line skates
    $8\frac{1}{4}$% sales tax ____________

Total Problems: **Total Correct:** **Score:**

Name ______________________________

Study the box below. Write each fraction as a percent. Round to the nearest hundredth. Write the answer on the line provided.

Rule:	**Example:**
To change the fraction to a percent: 1. Divide the numerator by the denominator. 2. Change the decimal to a percent. Move the decimal point 2 places to the right. Add a 0 if necessary, and do not forget the percent sign.	$\frac{7}{8}$ $7 \div 8 = .875$ **87.5%**

1. $\frac{11}{100}$ ______

2. $\frac{7}{14}$ ______

3. $\frac{5}{8}$ ______

4. $\frac{16}{25}$ ______

5. $\frac{1}{4}$ ______

6. $\frac{20}{120}$ ______

7. $\frac{1}{10}$ ______

8. $\frac{1}{3}$ ______

9. $\frac{106}{100}$ ______

10. $\frac{9}{20}$ ______

11. $\frac{5}{2}$ ______

12. $1\frac{1}{2}$ ______

Study the box below. Write each percent as a fraction or mixed number in lowest terms. Write the answer on the line provided.

Rule:	**Example:**
To change a percent to a fraction: 1. Drop the percent sign and place the number over 100. 2. Reduce the fraction to lowest terms.	$40\% = \frac{40}{100} = \mathbf{\frac{2}{5}}$

13. 30% = ______

14. 1% = ______

15. 75% = ______

16. 31% = ______

17. 67.5% = ______

18. 99% = ______

19. 0.3% = ______

20. 19% = ______

Total Problems:	**Total Correct:**	**Score:**

Name ________________________________

Study the box below. Express each decimal as a percent. Write the answer on the line provided.

Rule:	Example:
To change a decimal to a percent: Multiply by 100 (which moves the decimal point 2 places to the right) and add the percent sign.	0.7 0.7 x 100 = **70%**

1. 0.42 = ______
2. 0.4 = ______
3. 0.52 = ______
4. 0.25 = ______
5. 0.02 = ______
6. 0.05 = ______
7. 0.5 = ______
8. 0.92 = ______
9. 0.565 = ______
10. 0.29 = ______
11. 0.875 = ______
12. 1.42 = ______

Study the box below. Express each percent as a decimal. Write the answer on the line provided.

Rule:	Examples:	
To change a percent to a decimal: Divide by 100 (which moves the decimal point 2 places to the left).	71% 71 ÷ 100 = **.71**	8% 8 ÷ 100 = **.08**

13. 32% = ______
14. 4.5% = ______
15. 200% = ______
16. 61% = ______
17. 81.2% = ______
18. 0.1% = ______
19. 8% = ______
20. 70% = ______
21. 100% = ______
22. 1% = ______
23. 12% = ______
24. 0.05% = ______

Total Problems:	Total Correct:	Score:

Name ______________________________

Solve each word problem. Show your work and write the answer in the space provided.

1. Harrison earned $365 in commissions during the first week of the year by selling refrigerators. This was 18% of his total sales for the month. What were his total sales for the month?

2. Maria sold 28 dolls over the weekend at her toy store. This was 30% of all the dolls she had in stock. How many dolls were in stock?

3. Britt's team won 70% of their games. They played 20 games during the season. How many games did they win?

4. In the final game of Lana's season, 12 girls played. If 90% of the team played in that game, how many girls were on the team?

5. The cheerleaders ordered 250 spirit ribbons to sell at each basketball game. At the first game, they sold 179. What percent of the ribbons did they sell?

6. Refer to problem 5. During the second game, the cheerleaders sold 215 spirit ribbons. Find the percent of increase.

7. Cory traveled 18 miles in his car on 1 gallon of gasoline. At that same rate of fuel consumption, how far would he be able to go on 12 gallons?

8. Mollie was traveling in Israel on vacation. She purchased a souvenir for 65 shekels (Israeli currency). If the exchange rate was 3 shekels : 1 dollar, how many dollars did the souvenir cost?

Total Problems: **Total Correct:** **Score:**

Name ______________________________

Study the box below. On the line provided, write an algebraic expression or equation for each phrase.

Example:

Three times the cost of the sweater is $60.

Let c represent the cost of the sweater. The word "times" suggests multiplication, and "is" means an equal sign.

$3c = 60$

Tip:

Look for the key words which indicate addition, subtraction, multiplication, division, or equals.

1. the product of r and 9

   ____

2. 12 more than y

   ____

3. the quotient of w and 25

   ____

4. x subtracted from 21

   ____

5. m divided by 18

   ____

6. One-third of n is 12.

   ____

7. Twice a number is 28.

   ____

8. Two more than the quotient of 6 and x is 5.

   ____

9. Five increased by n is 10.

   ____

10. the sum of b and 16

    ____

11. the difference of y and 5

    ____

12. 16 less than e

    ____

13. the product of d and 15

    ____

14. Eleven times y is 22.

    ____

15. The sum of y and 5 is 32.

    ____

16. Eight decreased by a number is 3.

    ____

17. Fifteen decreased by the product of 9 and a number is 12.

    ____

18. Three less than the sum of 5 and y is 10.

    ____

Total Problems: **Total Correct:** **Score:**

Name ______________________________

Study the box below. Evaluate each expression if $a = 6$, $b = 4$, $c = 5$, and $d = 3$. Write the answer on the line provided.

Rule:	**Example:**
Substitute the numbers for the variables.	ad
Evaluate.	6 x 3 = **18**

1. b^2 ____________

2. abc ____________

3. $(abc)^2$ ____________

4. $3a - 5$ ____________

5. $a^2 - 2$ ____________

6. $7a - 2c$ ____________

7. $5b \div 2$ ____________

8. abc^2 ____________

9. $\frac{ab}{d} + c$ ____________

Evaluate each expression if $w = 9$, $x = 8$, $y = 4$, and $z = 2$. Write the answer on the line provided.

10. $x \div y$ ____________

11. $x^2 \div y^3$ ____________

12. $3xy^2$ ____________

13. $4x \div y^2$ ____________

14. $wx - yz$ ____________

15. $\frac{wx}{y}$ ____________

16. $3(xy)^2$ ____________

17. $3z^2$ ____________

18. $3w + x$ ____________

19. $2y^2 - z^2$ ____________

20. $2w \div z$ ____________

21. $wxyz^2$ ____________

Total Problems: **Total Correct:** **Score:**

Name ____________________

Solving Mental Math Equations

Study the box below. Solve each equation using mental math. Write the answer on the line provided.

Examples:

$2n = 18$

Think: What number times 2 equals 18?

$2 \times 9 = 18$
n = 9

$b - 12 = 5$

Think: What number minus 12 equals 5?

$17 - 12 = 5$
b = 17

1. $3x = 21$ ________
2. $y - 8 = 12$ ________
3. $\frac{y}{3} = 7$ ________
4. $c - 7 = 20$ ________
5. $7x = 49$ ________
6. $25 = 10 + x$ ________
7. $8y = 64$ ________
8. $9 = \frac{x}{9}$ ________
9. $\frac{x}{4} = 5$ ________
10. $y + 6 = 14$ ________
11. $n - 4 = 10$ ________
12. $\frac{n}{8} = 4$ ________
13. $0 = x - 9$ ________
14. $100 = 4y$ ________
15. $20 + n = 30$ ________
16. $50 = 30 + n$ ________
17. $m + 9 = 16$ ________
18. $6n = 36$ ________
19. $4x = 28$ ________
20. $x + 12 = 22$ ________
21. $n - 15 = 20$ ________
22. $\frac{x}{6} = 5$ ________
23. $4 = x - 8$ ________
24. $37 - x = 10$ ________

Total Problems: **Total Correct:** **Score:**

Name ______________________________

Study the box below. Solve and check each equation. Write the answer in the space provided.

Rule:	**Example:**
1. Look at what has been done to the variable.	$x - 144 = 120$
2. Undo it using the inverse (opposite) operation on both sides of the equation.	$x - 144 + 144 = 120 + 144$ $\mathbf{x = 264}$
3. To check, replace the variable with your solution to see if it makes the equation true.	Check: $264 - 144 = 120$ $120 = 120$

1. $42 + x = 66$

2. $c + 42 = 51$

3. $a + 4.9 = 7.6$

4. $z - 28 = 15.7$

5. $6\frac{2}{3} + y = 12\frac{1}{6}$

6. $y + 28 = 44$

7. $y - 46 = 27$

8. $e - 66 = 16$

9. $p + 2.6 = 12.2$

10. $m - 8.9 = 6.8$

11. $y - 145 = 207$

12. $x + 5\frac{1}{5} = 10\frac{1}{2}$

13. $45 + h = 83$

14. $n - 23 = 52$

15. $r - 0.4 = 11.3$

16. $n + 82 = 217$

17. $x - 73 = 291$

18. $y - 8\frac{2}{5} = 7\frac{2}{3}$

Total Problems: **Total Correct:** **Score:**

Name ______________________________

Solving Multiplication and Division Equations

Study the box below. Solve and check each equation. Write the answer in the space provided.

Rule:	**Example:**
1. Look at what has been done to the variable.	$3x = 21$
2. Undo it using the inverse (opposite) operation on both sides of the equation.	$3x \div 3 = 21 \div 3$ $\mathbf{x = 7}$
3. To check, replace the variable with your solution to see if it makes the equation true.	Check: $3 \times 7 = 21$ $21 = 21$

1. $8m = 72$

2. $c \div 6 = 9$

3. $\frac{h}{9} = 13$

4. $1.2x = 3.6$

5. $\frac{n}{3.6} = 0.8$

6. $n \div 12 = 4.8$

7. $\frac{1}{3}r = \frac{3}{4}$

8. $12x = 48$

9. $\frac{n}{14} = 18$

10. $16 = 0.4x$

11. $\frac{h}{1.5} = 0.3$

12. $15y = 195$

13. $\frac{1}{2}c = \frac{2}{5}$

14. $n \div 8 = 92$

15. $\frac{n}{12} = 15$

16. $\frac{x}{2.4} = 9.5$

17. $11.52 = 4.8v$

18. $\frac{4}{5}n = \frac{1}{2}$

Total Problems: **Total Correct:** **Score:**

Name ___________________________________

Study the box below. Solve and check the following equations. Write the answer in the space provided.

Rules:	Example:
1. To solve a 2-step equation, first undo the addition and subtraction. 2. Then, undo the multiplication and division. 3. To check, replace the variable with your solution to see if it makes the equation true.	$4x + 5 = 17$ $4x + 5 - 5 = 17 - 5$ $4x = 12$ $4x \div 4 = 12 \div 4$ $\mathbf{x = 3}$ Check: $4 \bullet 3 + 5 = 17$ $12 + 5 = 17$ $17 = 17$

1. $3y + 6 = 27$

2. $5y - 2.6 = 7.4$

3. $6p + 14 = 68$

4. $\frac{h}{12} + 7 = 12$

5. $12x - 9 = 75$

6. $\frac{n}{7.2} + 4.5 = 6.4$

7. $\frac{x}{7} + 5 = 10$

8. $8c + 9 = 65$

9. $\frac{x}{15} - 6.3 = 2.7$

10. $\frac{y}{8} + 19 = 28$

11. $6.2w - 2.3 = 25.6$

12. $\frac{b}{9} - 12 = 8$

13. $42 = 12r - 6$

14. $16m - 14 = 34$

15. $27c - 18 = 36$

Total Problems: **Total Correct:** **Score:**

Name ______________________________

Solve and check the following equations. Write the answer in the space provided.

1. $x - 14 = {}^{-}38$
2. $y - 50 = {}^{-}20$
3. ${}^{-}20 = v + 26$
4. ${}^{-}5x - 7 = 28$
5. $3(x - 5) = {}^{-}27$
6. $\frac{x}{3} - 12 = {}^{-}20$
7. ${}^{-}3x - 5 = 13$
8. $5y = {}^{-}35$
9. $\frac{y}{3} = {}^{-}9$
10. $3(x - 2) = {}^{-}9$
11. $3y + 6 = {}^{-}3$
12. $2(h + 3) = {}^{-}6$
13. $\frac{a}{2} + 4 = {}^{-}10$
14. ${}^{-}9x + 3 = 30$
15. $x + 150 = {}^{-}125$
16. $\frac{m}{{}^{-}6} = {}^{-}13$
17. ${}^{-}4x = 16$
18. ${}^{-}2x + 58 = {}^{-}8$
19. ${}^{-}3x - 1 = {}^{-}10$
20. ${}^{-}30 = \frac{m}{5} - 15$

Total Problems: **Total Correct:** **Score:**

Name ______________________________

Solve each word problem. Show your work and write the answer in the space provided.

1. The phrase "6 less than x" can be represented by the expression x– 6. Write 2 additional phrases that could be represented by x – 6.

2. Callie went berry picking over the weekend. She picked 80 berries. She wants to divide them equally among her 4 teachers. How many berries will Callie give to each teacher?

3. If x is greater than 1, would 2x be greater than, less than, or equal to 2? Give an example to support your answer.

4. Visualize a balance scale with an equal number of 1-gram weights on each pan. What would happen to the scale if 2 weights were removed from the left side? What can be done to the right side of the scale to put it back into balance?

5. A new bike was on sale for $25 off the original price. If the sale price was $180, write and solve an equation for the original price of the bike.

6. Yakim works on a police force with 48 other officers. During a downtown parade, each officer is told to cover 200 feet. What is the total number of feet to be covered during the parade?

7. Rory has 5 times as many baseball cards as Billy. If Rory has 60 baseball cards, how many does Billy have?

8. Of the 128 guests at Iris's wedding, 75 ordered chicken, 46 ordered beef, and the rest requested vegetarian meals. Write and solve an equation which shows how many people requested a vegetarian meal.

Total Problems: **Total Correct:** **Score:**

Name ______________________________

Study the rules below. Classify each angle as acute, right, obtuse, or straight. Write the answer in the space provided. Draw the angle, if necessary.

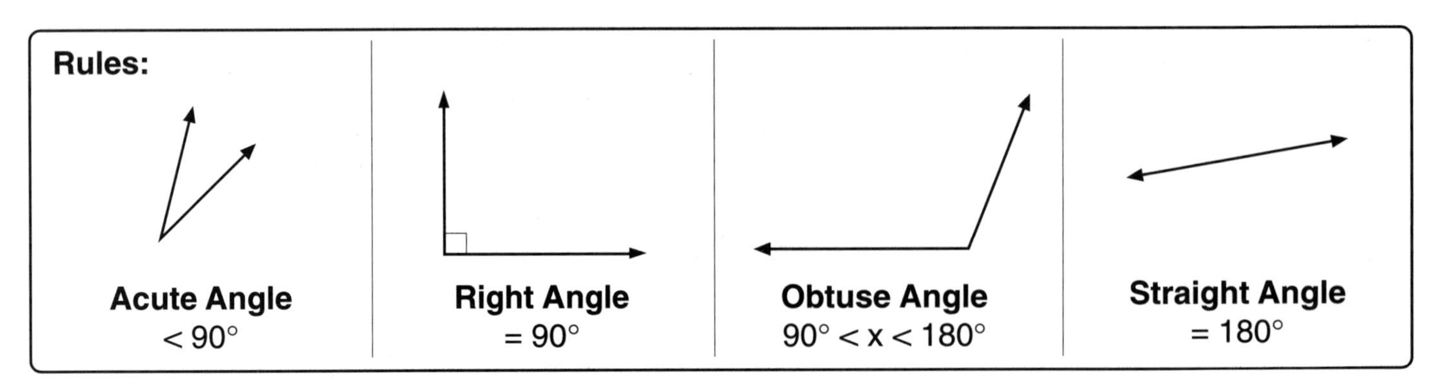

1.

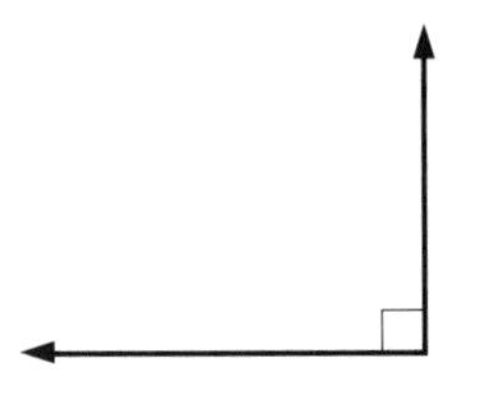

2.

3.

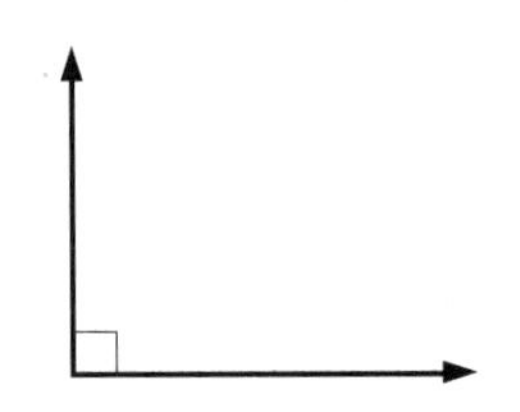

4. 115°

5.

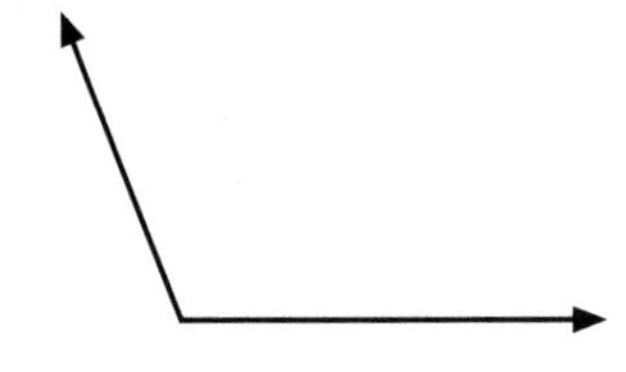

6.

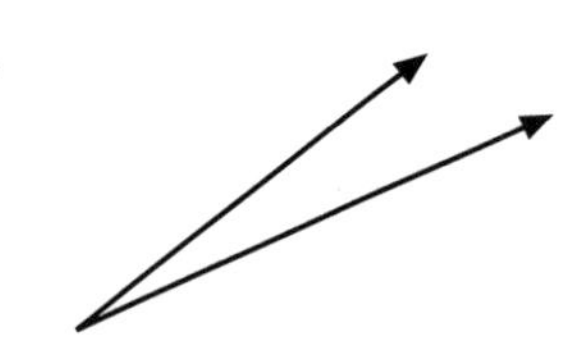

7.

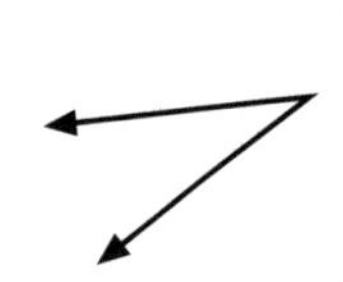

8. 90°

9.

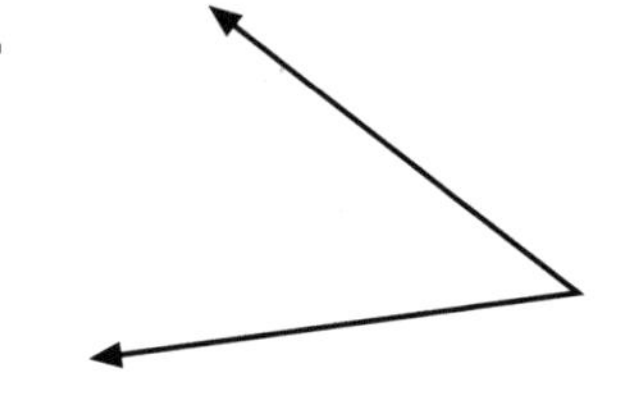

10.

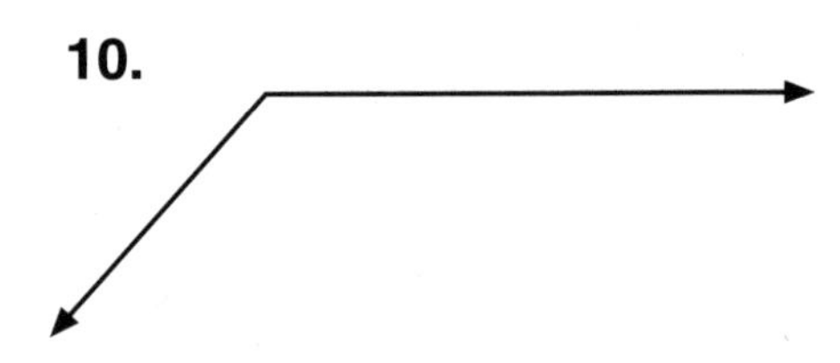

11.

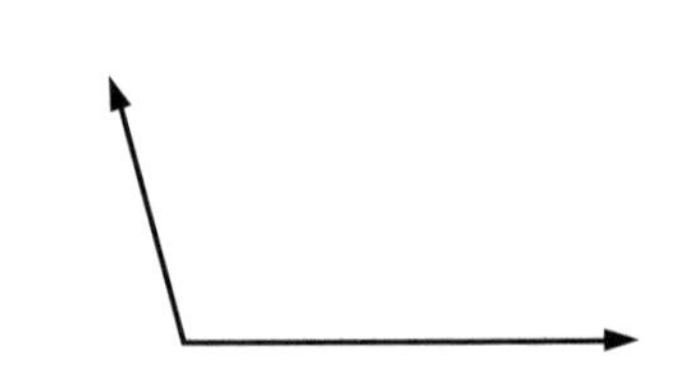

12. 43°

Total Problems: Total Correct: Score:

Name ______________________________

Study the rules below. Classify each triangle by its sides and angles.

Rules:	**Equilateral**	**Isosceles**	**Scalene**
By sides:	 3 equal sides	2 equal sides	no equal sides
By angles:	**Acute** 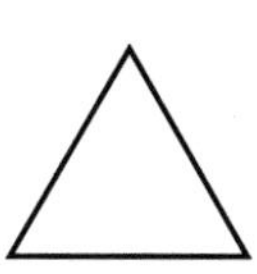 3 acute angles	**Obtuse** 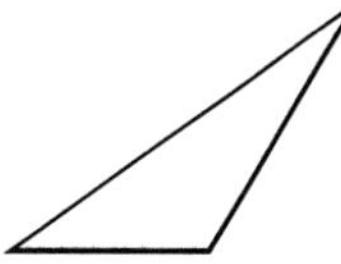 1 obtuse angle	**Right** 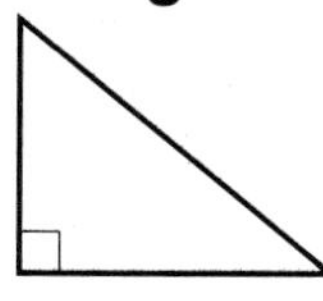 1 right angle

1.

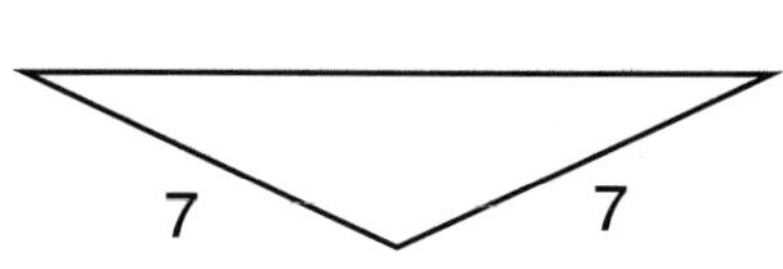

____

____

2.

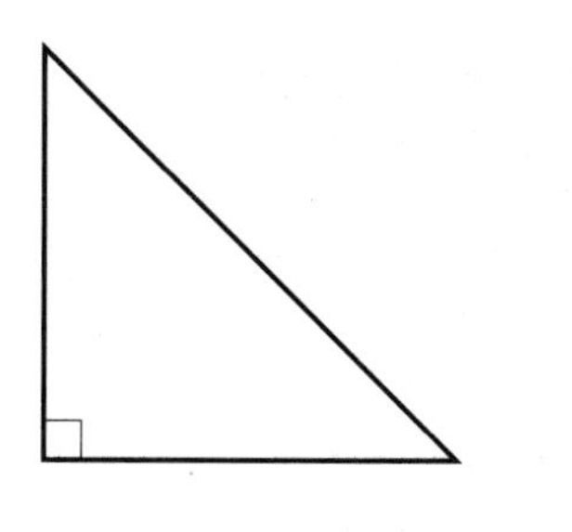

____

____

3.

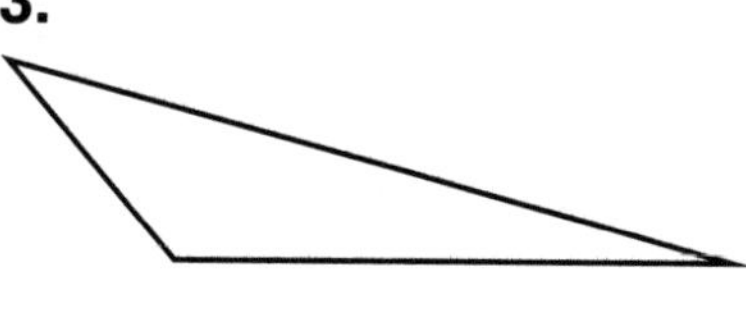

____

____

4.

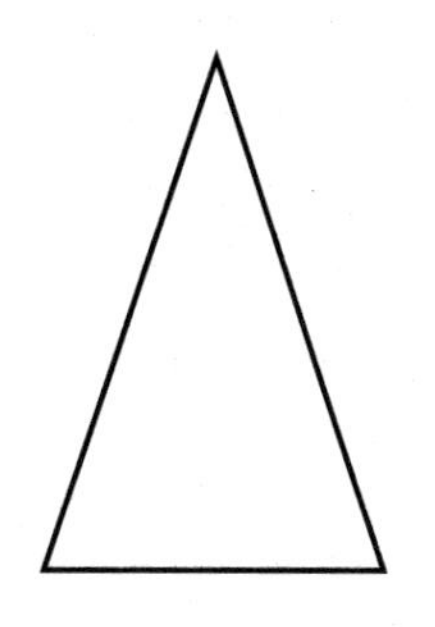

____

____

5.

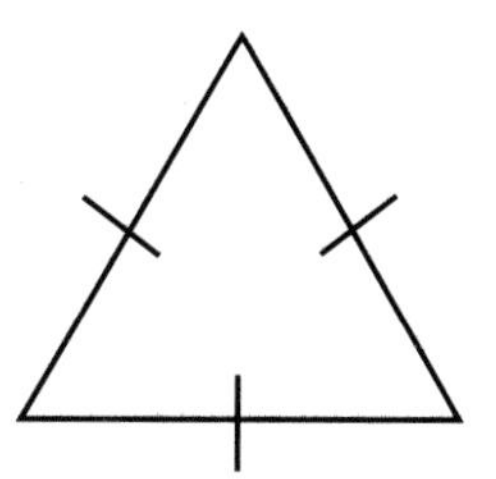

____

____

6.

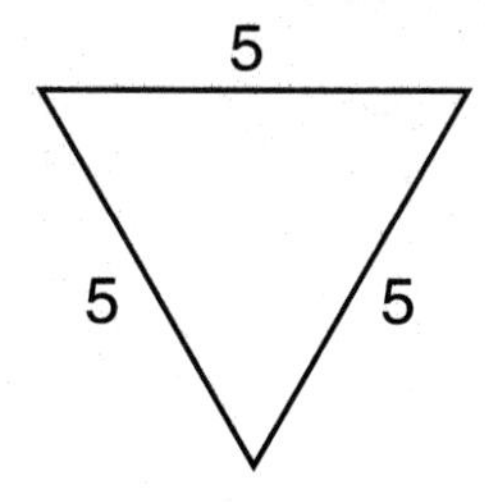

____

____

Total Problems: **Total Correct:** **Score:**

Name ______________________________

Study the rules below. Classify each quadrilateral as a square, rectangle, rhombus, parallelogram, or trapezoid. (Some quadrilaterals may have more than one name.) Write the answer(s) on the lines provided.

Rules:	Square	Parallelogram	Rectangle	Rhombus	Trapezoid
	All sides equal All angles 90°	Opposite sides parallel	Opposite sides equal All angles 90°	All sides equal Opposite angles equal	1 pair of parallel sides

1.

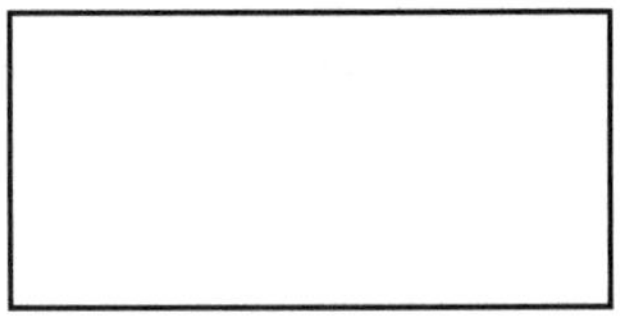

____

____

2.

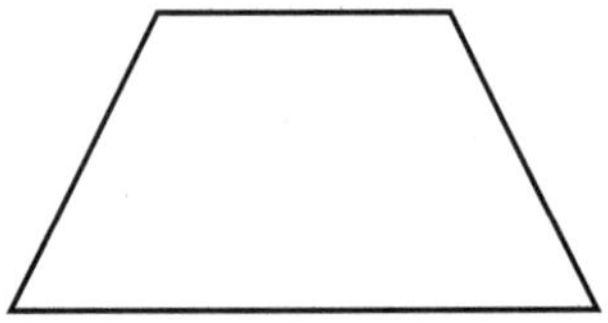

____

____

3.

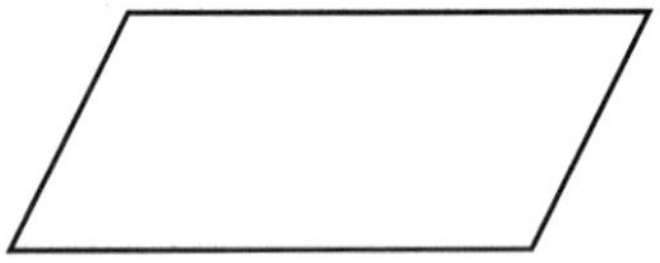

____

____

4.

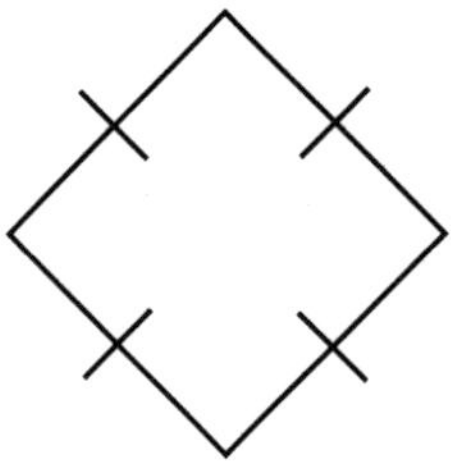

____

____

5.

____

____

6.

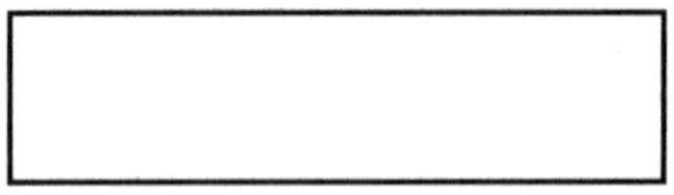

____

____

7.

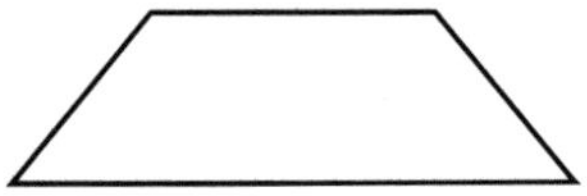

____

____

8.

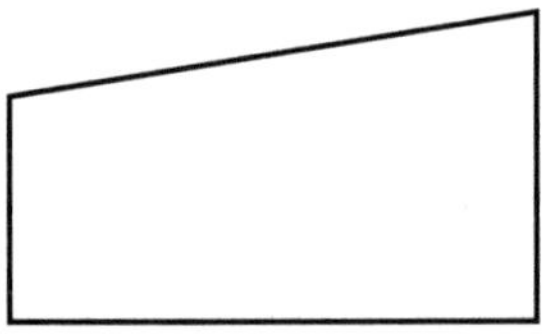

____

____

9.

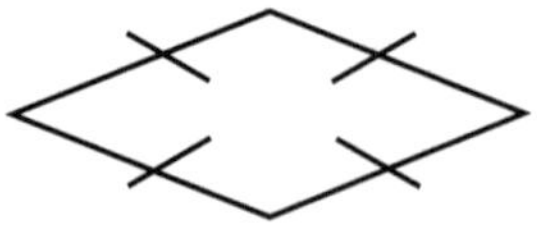

____

____

Total Problems: **Total Correct:** **Score:**

Name ______________________________

Study the box below. Find the area and perimeter of each figure. Write the answers on the lines provided.

Rules:

Area: the measure of a region expressed in square units
Perimeter: distance around a region expressed in units

Examples:

Square

5 in

A = side^2
= 5^2
= **25 in^2**
P = s x 4
= 5 x 4
= **20 in**

Rectangle

6 m

21 m

A = length x width
= 21 x 6
= **126 m^2**
P = (2 x l) + (2 x w)
= (2 x 21) + (2 x 6)
= 42 + 12
= **54 m**

Parallelogram

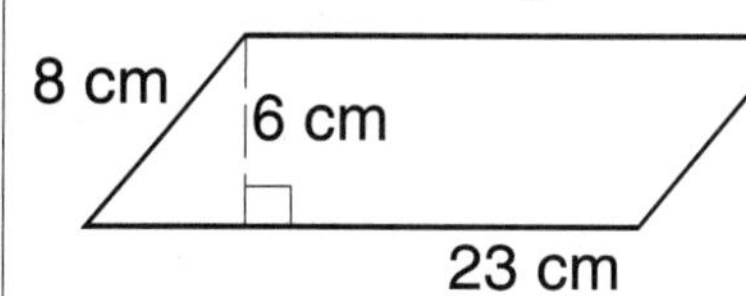

A = base x height
= 23 x 6
= **138 cm^2**
P = (2 x l) + (2 x w)
= (2 x 23) + (2 x 8)
= 46 + 16
= **62 cm**

1.

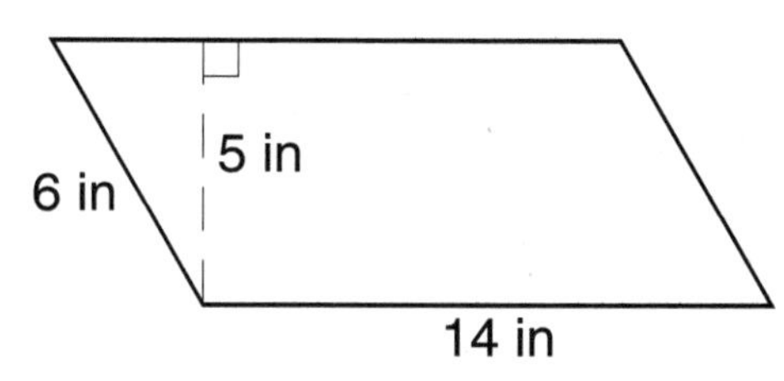

A = ________

P = ________

2.

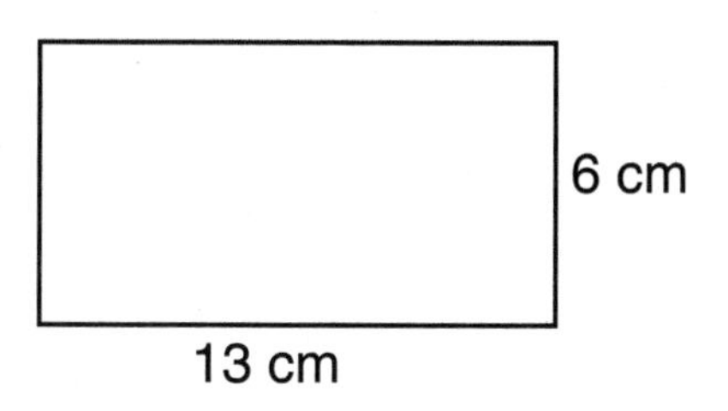

A = ________

P = ________

3.

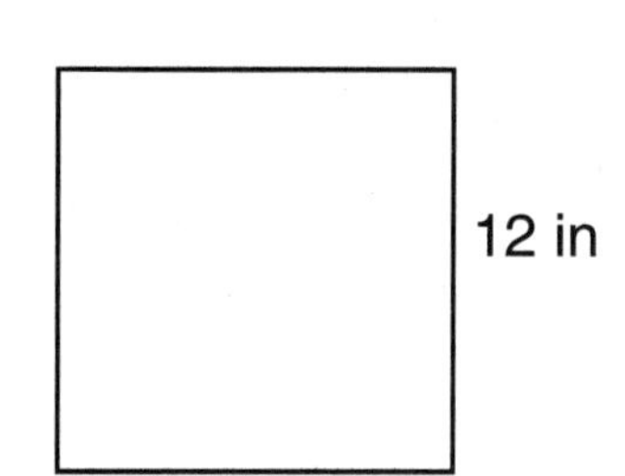

A = ________

P = ________

4.

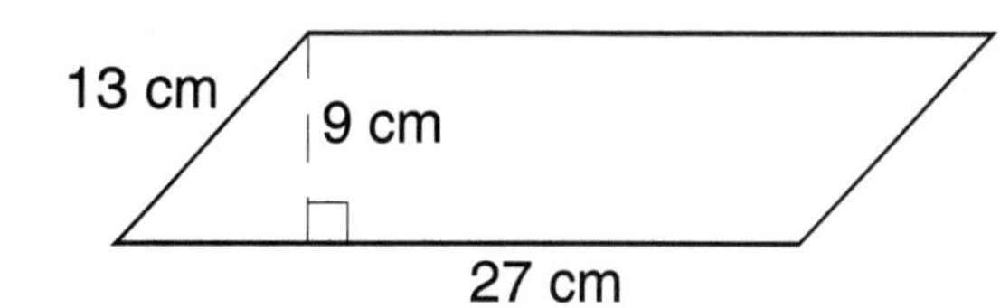

A = ________

P = ________

Total Problems: **Total Correct:** **Score:**

Name ____________________

Study the box below. Find the area of each triangle. Write the answer on the line provided.

Rule:

Area: the measure of a region expressed in square units.

The area of a triangle = $\frac{1}{2}$(base x height).

The base and height of a triangle will always be perpendicular.

Example:

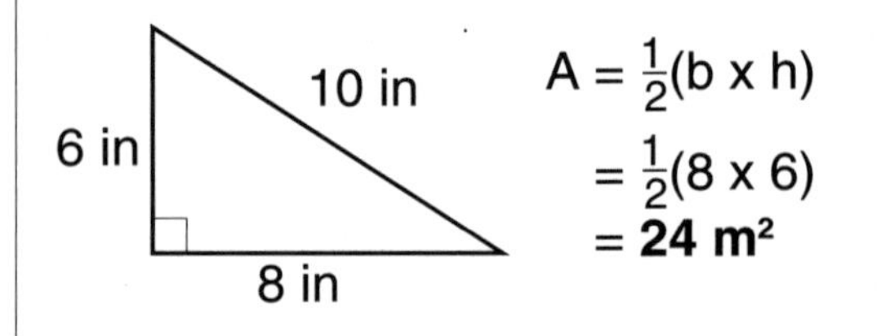

$A = \frac{1}{2}(b \times h)$

$= \frac{1}{2}(8 \times 6)$

$= \mathbf{24\ m^2}$

1.

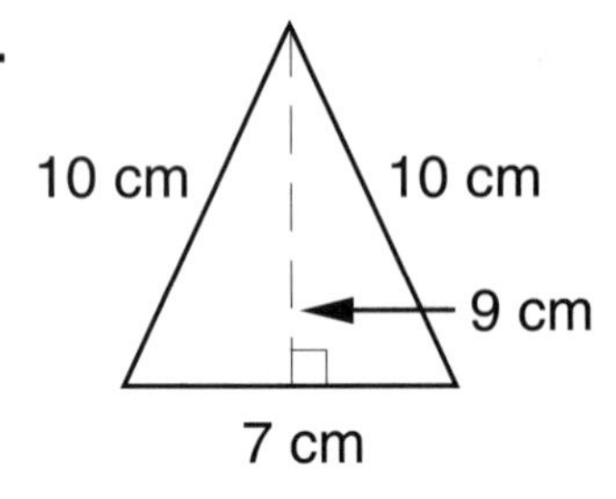

____

2.

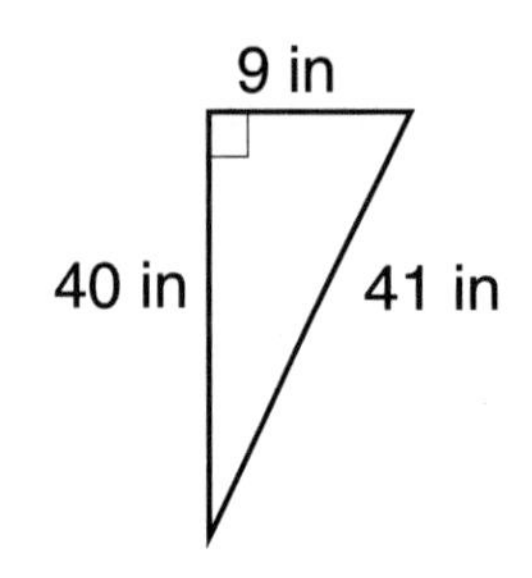

____

3.

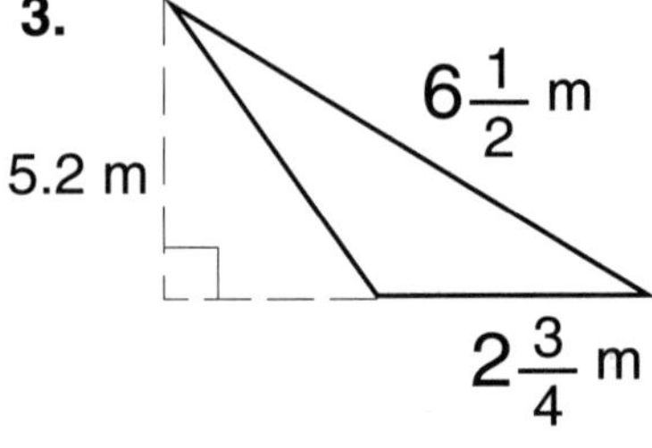

____

Study the box below. Find the area of each trapezoid. Write the answer on the line provided.

Rule:

The area of a trapezoid = $\frac{1}{2}h(a + b)$, where h is the height and a and b are the bases.

Example:

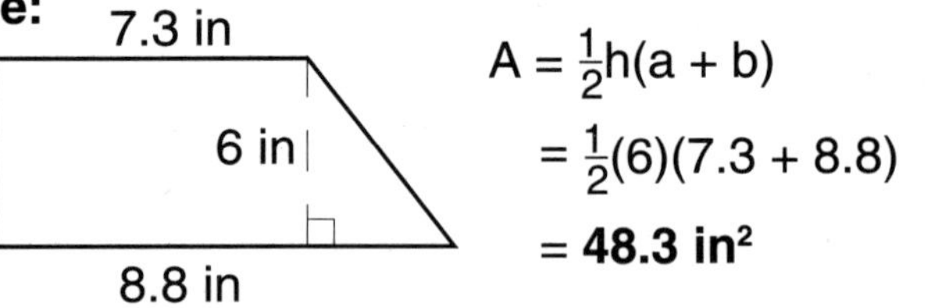

$A = \frac{1}{2}h(a + b)$

$= \frac{1}{2}(6)(7.3 + 8.8)$

$= \mathbf{48.3\ in^2}$

4.

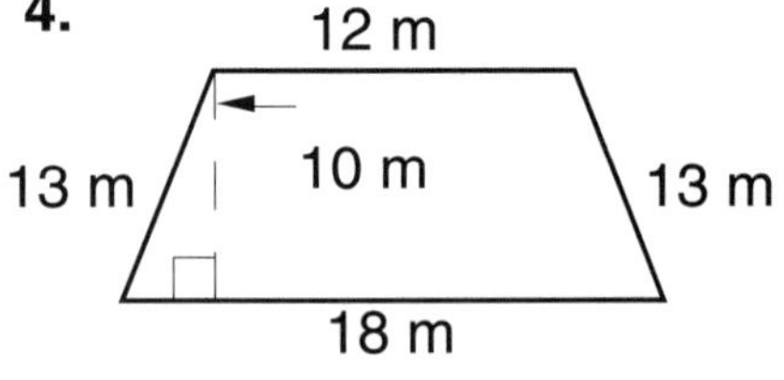

____

5.

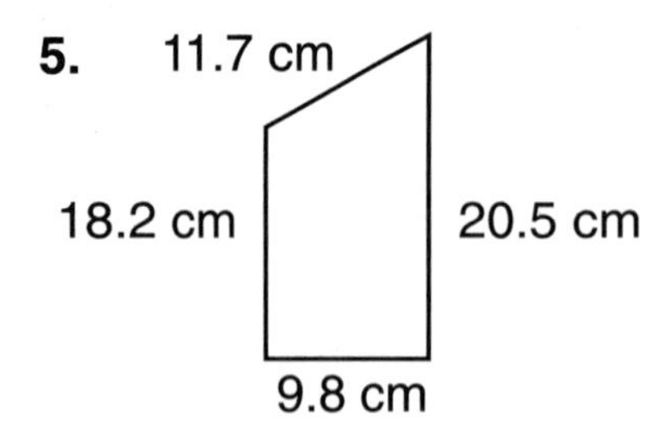

____

6. bases: 6.3 m, 8.5 m
height: 7.8 m

____

Total Problems: **Total Correct:** **Score:**

Name ______________________________

Study the box below. Find the circumference and area of each circle. Use 3.14 for π. Round to the nearest hundredth. Write the answers on the lines provided.

Rules:

Circumference: distance around the circle
$C = \pi(\text{diameter})$

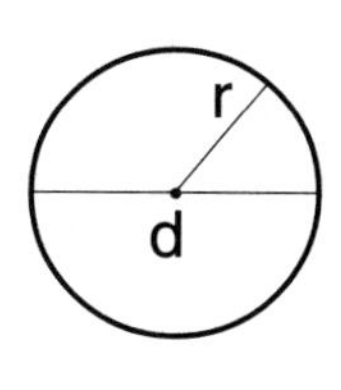

Area: measure of the region inside the circle
$A = \pi(\text{radius})^2$

$r = \frac{1}{2}d$
$d = 2r$

Example:

16 ft

$C = \pi d$
$= 3.14 \times 16$
$= \mathbf{50.24\ ft}$

$A = \pi r^2$
$= 3.14 \times 8^2$
$= \mathbf{200.96\ ft^2}$

1.

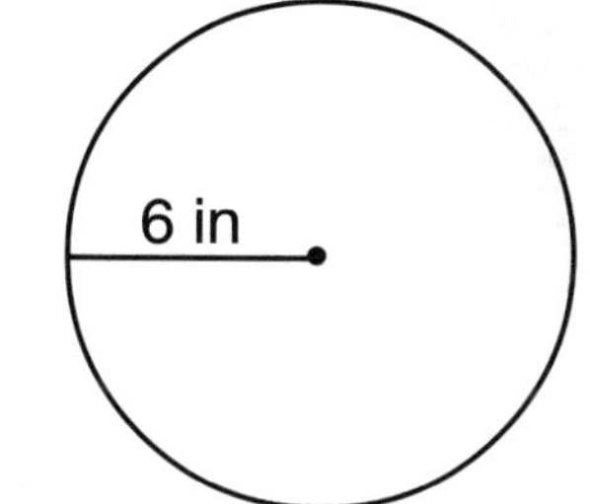

C = ______
A = ______

2.

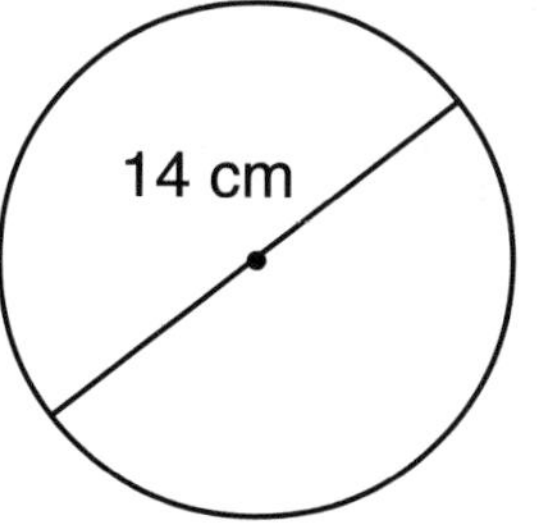

C = ______
A = ______

3.

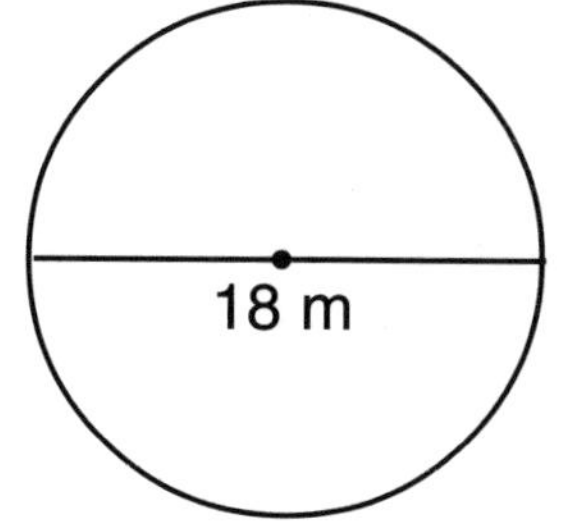

C = ______
A = ______

4. d = 10.5 km

C = ______
A = ______

5.

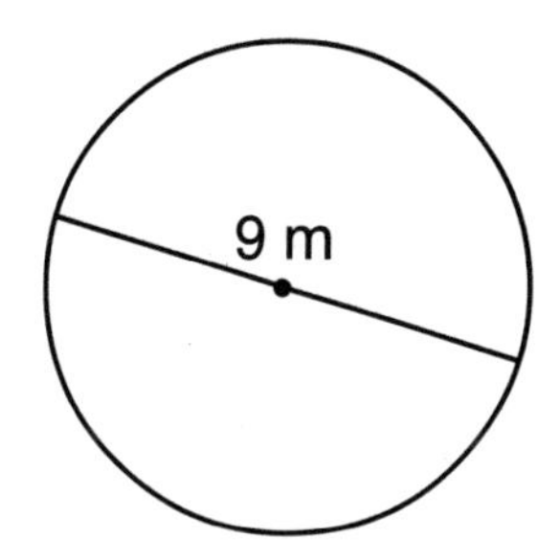

C = ______
A = ______

6.

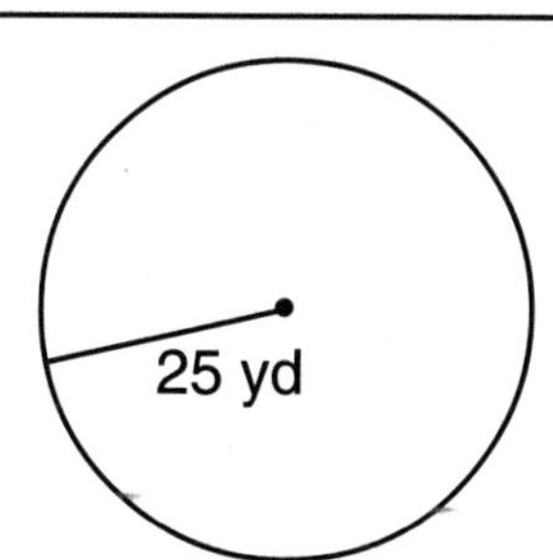

C = ______
A = ______

7.

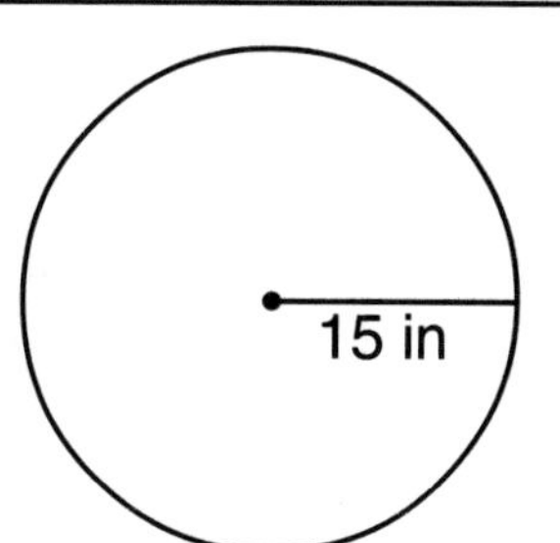

C = ______
A = ______

8. r = 8.1 ft

C = ______
A = ______

Total Problems: Total Correct: Score:

Name ______________________________

Use the coordinate system below to identify the coordinates of each point. The first coordinate is the distance from 0 on the x-axis, and the second coordinate is the distance from 0 on the y-axis.

1. A 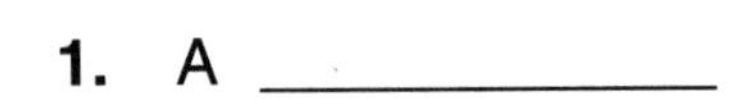

2. B 

3. C 

4. D 

5. E 

6. F ______________

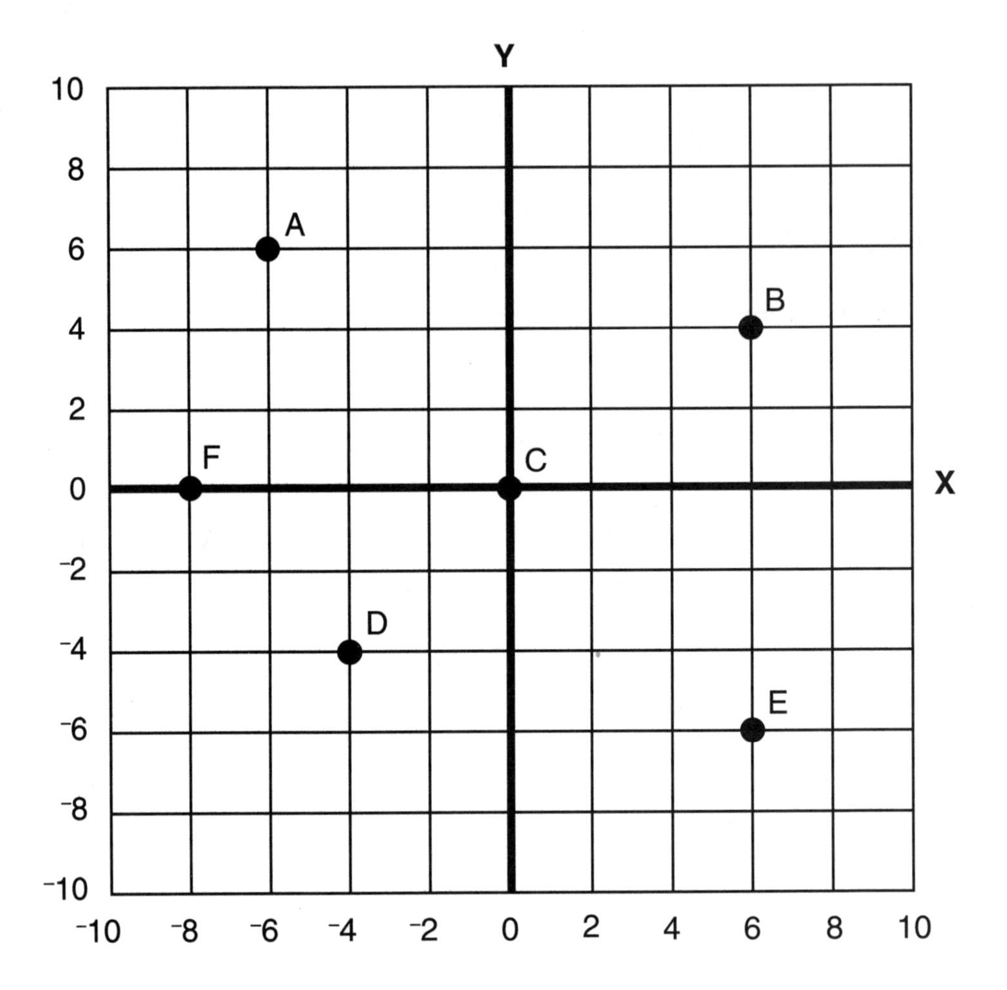

On a sheet of graph paper, draw a coordinate plane. Label the x- and y-axis. Then, graph each set of points and label each point.

7. Q $(^-4, ^-1)$
8. R $(2, 5)$
9. S $(3, ^-6)$
10. T $(7, 0)$
11. U $(^-5, 2)$
12. V $(^-2, ^-4)$
13. W $(0, 1)$
14. X $(0, 0)$
15. Y $(^-3, 0)$
16. Z $(0, ^-3)$

Total Problems: Total Correct: Score:

Name ______________________________

Solve each word problem. Show your work and write the answer in the space provided.

1. To bisect an angle means to cut it in half. Write which type of angle is formed if each of the following angles are bisected:

A. acute

B. right

C. obtuse

D. straight

2. Is it possible for a trapezoid to have 2 equal sides? Draw a figure to support your answer.

3. Which quadrilaterals have 4 congruent sides?

4. Find the area and circumference of a circle whose radius is 12 meters. What would happen to its area and circumference if you were to double its radius?

5. The area of a parallelogram is bh. The area of a triangle is $\frac{1}{2}$bh. Write 2 sentences describing their relationship. (bh = base x height)

6. Is a square also a rectangle? Is it a parallelogram? Explain your answer.

Total Problems: **Total Correct:** **Score:**

Name ______________________________

Use the bar graph to answer the questions. Circle the letter beside the correct answer or write the answer on the lines provided.

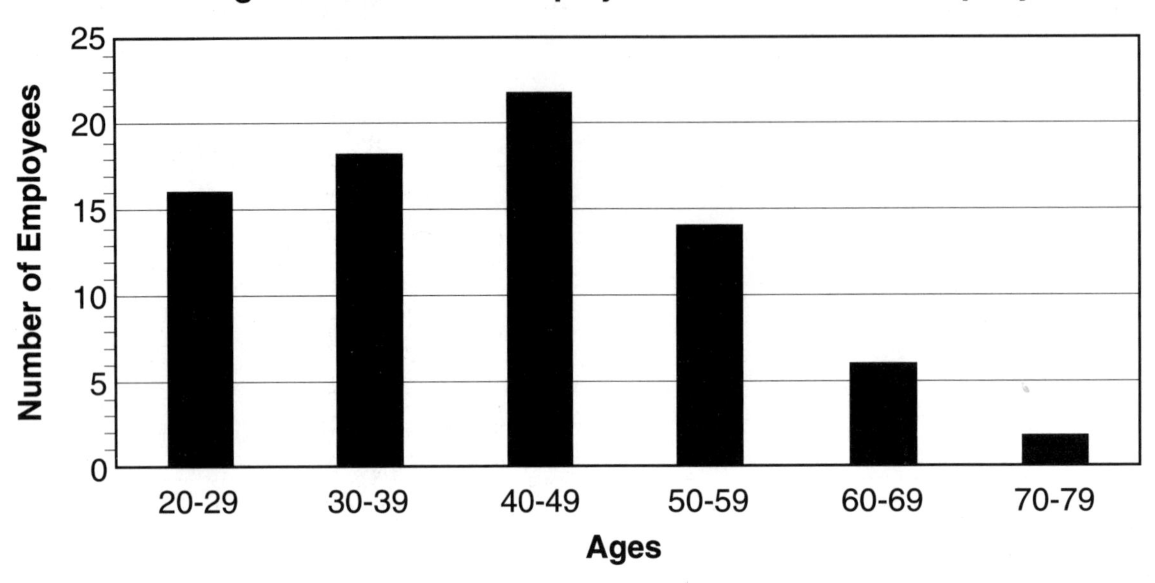

1. How many employees are between the ages of 40 and 49?
   A. 18 B. 22 C. 24 D. 16
2. How many more employees are between the ages of 50-59 than 60-69?
   A. 12 B. 10 C. 8 D. 5
3. How many employees work at The John's Company?
   A. 72 B. 78 C. 82 D. 88
4. Which age group has exactly 14 employees?
   A. 20-29 B. 30-39 C. 40-49 D. 50-59
5. How many employees are under age 40?
   A. 39 B. 21 C. 30 D. 34
6. How many more employees are between the ages of 40-49 than 20-29?
   A. 10 B. 6 C. 22 D. 2
7. Why do you think the oldest age group, 70-79, has only 2 employees?

____

____

____

Total Problems: **Total Correct:** **Score:**

Name ______________________________

Use the line graph to answer the questions. Circle the letter beside the correct answer or write the answer on the lines provided.

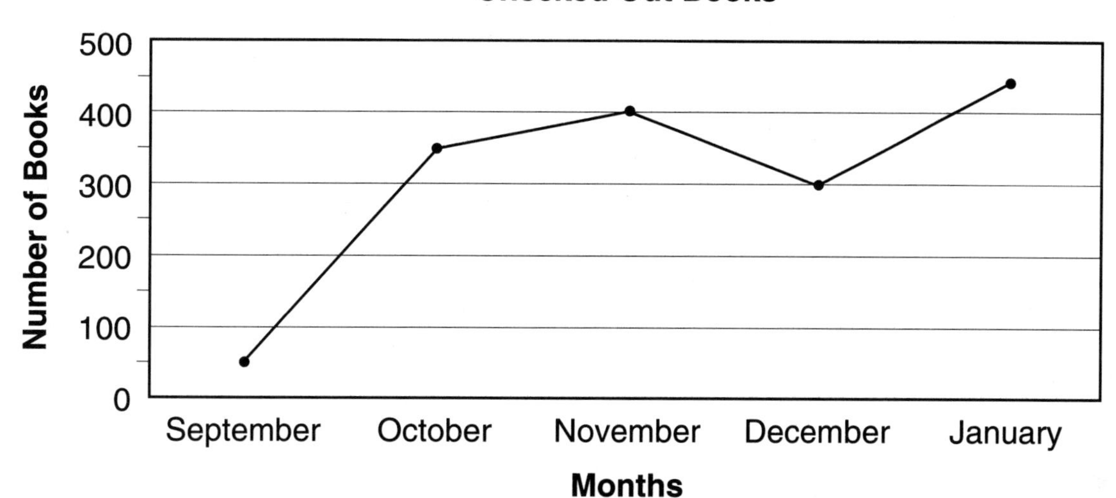

1. How many books were checked out during the month of October?
   A. 300 B. 350 C. 400 D. 450

2. How many more books were checked out in January than December?
   A. 150 B. 175 C. 200 D. 225

3. During which month were 300 books checked out?
   A. October B. November C. December D. January

4. Why do you think so few books were checked out in September?

5. The librarian would like an average of 350 books checked out per month. Has her goal been reached so far? Support your answer.

6. If there were 1,700 books in the library, what percentage of books were checked out in November?

Total Problems: **Total Correct:** **Score:**

Name ______________________________

Use the tables to answer the questions. Write the answer on the line provided.

1. What is the greatest number of books a student has read?

____

2. Who has read the fewest number of books?

____

3. What was the average number of books read by each student?

____

Mrs. Weber's Class

Number of Students	Number of Books Read
Marti	7
Terry	12
Rachel	9
Miquel	19
Kerren	8
Ernie	4
Stephanie	16

---

4. Which player scored the most points?

____

5. If the opposing team scored 61 points, did the Knights win or lose the game?

____

Knights Basketball Team

Player	Points
Perez	9
Johnson	8
Duval	17
Blanchard	25
Schultz	16

---

6. How many red pairs of shoes were sold?

____

7. How many more white, red, and blue shoes were sold than black?

____

8. Which shoe color should the store order the most? Justify your answer.

____

____

Color of Shoes Sold in One Hour

Color	Pairs Sold
Black	12
White	9
Red	3
Blue	6

Total Problems: **Total Correct:** **Score:**

Name ______________________________

Use the data in each table to make a line graph. Draw the graph in the space provided.

1. **Growth of Mindy's Baby**

Day	Weight (in Pounds)
July 1	12
August 1	14
September 1	17
October 1	18
November 1	20

2. **Monthly Electric Bill**

Month	Amount
January	$108
February	$110
March	$95
April	$63
May	$48

Total Problems: **Total Correct:** **Score:**

Name ______________________________

Study the box below. In the space provided, make a stem and leaf plot for each set of data.

Rule:

In a stem and leaf plot, the digits to the left of the line have the greater place value and are called stems. Digits to the right of the line represent digits in the ones place and are called leaves.

Example: 41, 42, 50, 39, 53 48, 51, 38, 36, 48

stem	leaf
3	689
4	1288
5	013

1. 22, 21, 32, 42, 45, 55, 56, 24, 24, 26

2. 89, 87, 99, 100, 95, 72, 78, 88, 88, 97, 98

3. 210, 215, 217, 208, 229, 225, 216, 225, 205

Use the following stem and leaf plot to answer the questions.

Height (in inches) of the basketball team

stem	leaf
6	168
7	001257
8	0

4. What is the height, in feet, of the shortest player? the tallest? ______________________

5. Which height is the most frequent? ______________________

6. How many people are on the team? ______________________

7. Where do most of the heights fall? ______________________

8. Is there a player who is exactly 7 feet tall? ______________________

Total Problems: **Total Correct:** **Score:**

Name ______________________________

Study the box below. Find the mean, median, and mode of each set of data. Round to the nearest tenth. Write the answers in the space provided.

Rules:	**Example:**
Mean (average): Add the numbers and divide by the total number in the set. **Median** (middle number): Place the numbers in order. Find the middle number. If there is not one middle number, average the two in the middle. **Mode** (most frequent): Find the number that occurs most frequently.	17, 15, 35, 10, 21, 11 Mean: (17 + 15 + 35 + 10 + 21 + 11) ÷ 6 = 109 109 ÷ 6 = 18.16 Rounds to **18.2** Median: The two middle numbers are 15 and 17. Reorder numbers: 10, 11, 15, 17, 21, 35 (15 + 17) ÷ 2 = **16** Mode: **No mode**

1. 30, 42, 50, 51, 40, 45, 50

mean: ________
median: ________
mode: ________

2. 65, 65, 90, 60, 80, 62, 80, 62, 62

mean: ________
median: ________
mode: ________

3. 34, 33, 39, 37, 29, 31, 36, 34

mean: ________
median: ________
mode: ________

4. 235, 245, 330, 235, 320, 325, 435

mean: ________
median: ________
mode: ________

Tell whether the mean, median, or mode would be the best measure for each given situation. Explain your answer.

5. Would you use mean, median, or mode to describe the typical selling price of a bicycle?

____

____

6. Would you use mean, median, or mode to determine the most popular toy sold at a store?

____

____

Total Problems:	Total Correct:	Score:

Name ______________________________

Study the box below. Find the total number of outcomes in each situation. Show your work and write the answer in the space provided.

Rule:	**Example:**
Using the Fundamental Counting Principle, multiply the number of choices in each set to derive the number of possible combinations.	The corner restaurant offers 4 different kinds of soups and 7 different kinds of sandwiches. How many soup/sandwich combinations are possible? 4 soups x 7 sandwiches = **28 combinations**

1. Tossing a quarter and rolling a number cube

2. Choosing from 5 different kinds of shirts and 4 different kinds of pants

3. Choosing a car in 1 of 5 colors, a dark or light interior, with either a cassette player or CD player

4. Choosing a ham, turkey, bologna, or salami sandwich with chips, crackers, or pretzels

5. Choosing 1 appetizer from 7 choices, 1 entree from 9 choices, and 1 dessert from 5 choices

6. Choosing from 4 history courses, 3 science courses, 2 math courses, and 3 English courses

Total Problems: **Total Correct:** **Score:**

Name ____________________

Study the box below. Solve each problem in the space provided.

Rule:	**Example:**
Experimental Probability: The probability based on the outcomes of an experiment. **Theoretical Probability:** The probability based on mathematic principles.	Find the theoretical probability of a number cube landing on two. $P(2) = \frac{1}{6}$ ← number of twos on the number cube; ← number of possible outcomes

1. Find the theoretical probability of choosing a boy to participate in a ceremony from 12 boys and 14 girls.

2. Mabel tosses a coin 100 times. It lands on heads 57 times.

A. What is the experimental probability of getting heads? tails?

B. Explain how the theoretical probability of getting heads compares with the experimental probability.

3. Find the theoretical probability of:

A. Rolling a sum of 2 on 2 dice.

B. Rolling a sum greater than 7 on 2 dice.

4. Matthew has a bag with 3 blue marbles, 4 green, 5 red, and 2 white.

A. What is the theoretical probability of choosing a white marble?

B. What is the theoretical probability of choosing a red or blue marble?

C. What is the theoretical probability of not choosing a green marble?

Total Problems: **Total Correct:** **Score:**

Name ______________________________

Study the box below. Make a tree diagram to show all the outcomes for each situation. Then, give the total number of outcomes. Show your work in the space provided.

Rule:	**Example:**
A tree diagram is used to show the total number of possible outcomes in a probability experiment.	Flipping 2 coins. Coin 1 — H / T Coin 2 — H / T There are 4 possible outcomes: HH, HT, TH, TT.

1. Choosing a small, medium, or large T-shirt in white, red, or black

2. Choosing a letter A, B, or C, and choosing a number 1, 2, or 3

3. Choosing French toast or pancakes and milk or orange juice

4. Choosing a navy, white, or burgundy suit with pants or a skirt

Total Problems: **Total Correct:** **Score:**

Name ________________________________

Study the box below. Find the number of combinations. Write the answer in the space provided.

Rule:

A **combination** is an arrangement of items in which order does not matter.

A **permutation** is an arrangement of items in a particular order.

Example:

5 players from 9 — If you have 9 players and only 5 can play at a time, how many different combinations of 5 players are there?

$$\frac{9 \bullet 8 \bullet 7 \bullet 6 \bullet 5}{5 \bullet 4 \bullet 3 \bullet 2 \bullet 1} = \frac{15{,}120}{120}$$

15,120 ← Number of permutations of 9 players taken 5 at a time

120 ← Number of permutations of 5 players

$$\frac{15{,}120}{120} = \mathbf{126}$$

There are 126 combinations of players.

1. 4 movies from a list of 7

2. 4 numbers from the numbers 1-5

3. 3 numbers from 5, 10, 15, 20

4. 2 letters from A, B, C, D, and E

5. 4 girls from a group of 8

6. 2 bows from 6 bows

7. 5 shirts out of 12 shirts

8. 3 applications from a choice of 15

Total Problems: **Total Correct:** **Score:**

Name ______________________________

Solve each word problem. Show your work and write the answer in the space provided.

1. Suppose your teacher gives you a 4-question true/false quiz. How many sets of guesses are possible?

2. Mark is choosing from among 4 brands of golf clubs, 2 kinds of golf balls, and 3 types of golf bags. In how many different ways can he buy a set of golf clubs, golf balls, and a golf bag?

3. A soft drink vending machine contains 2 soda buttons, 2 diet soda buttons, a ginger ale button, a root beer button, and an orange soda button. If you were to randomly choose a drink without looking, what would be the theoretical probability of choosing a diet soda?

4. In how many ways can a baseball coach choose 3 pitchers from a group of 5 able pitchers?

5. During the first semester of Camille's literature class, each student must read and write about a short story, poem, and play. If there are 27 short stories, 18 poems, and 2 plays in her literature book, how many different selections can be made?

6. Steph decides to personalize her license plate. If the first 3 characters must be numbers between 1 and 9 (without repeating) and the next 3 characters must be a letter (any letter A-Z without repeating), how many possible ways can she create her license plate?

Total Problems: **Total Correct:** **Score:**

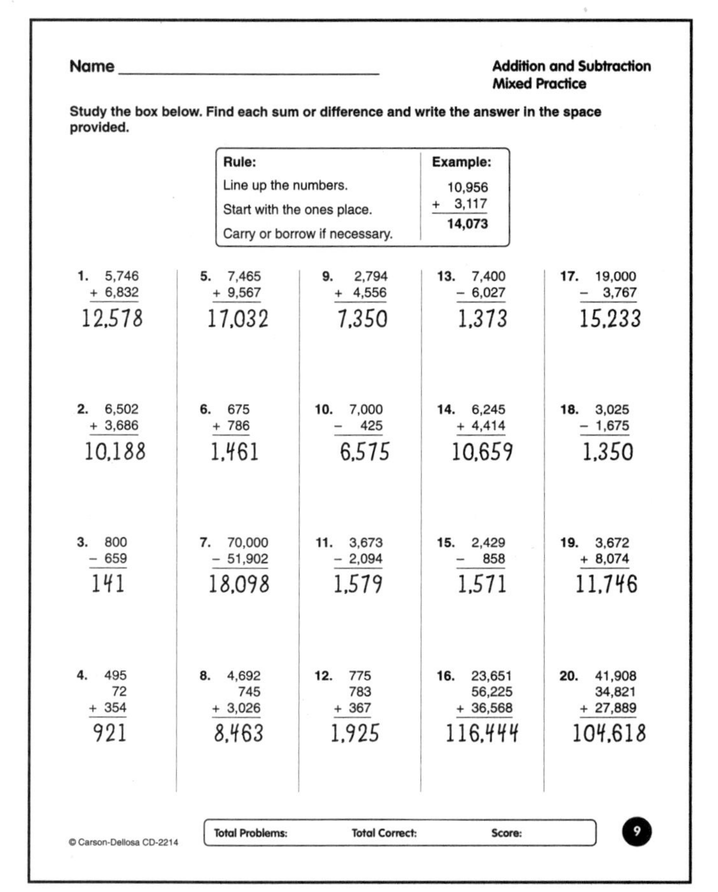

Name ____________________ **Addition and Subtraction Mixed Practice**

Study the box below. Find each sum or difference and write the answer in the space provided.

Rule:	Example:
Line up the numbers. Start with the ones place. Carry or borrow if necessary.	10,956 + 3,117 **14,073**

1. 5,746 + 6,832 = 12,578
2. 6,502 + 3,686 = 10,188
3. 800 − 659 = 141
4. 495 + 72 + 354 = 921
5. 7,465 + 9,567 = 17,032
6. 675 + 786 = 1,461
7. 70,000 − 51,902 = 18,098
8. 4,692 + 745 + 3,026 = 8,463
9. 2,794 + 4,556 = 7,350
10. 7,000 − 425 = 6,575
11. 3,673 − 2,094 = 1,579
12. 775 + 783 + 367 = 1,925
13. 7,400 − 6,027 = 1,373
14. 6,245 + 4,414 = 10,659
15. 2,429 − 858 = 1,571
16. 23,651 + 56,225 + 36,568 = 116,444
17. 19,000 − 3,767 = 15,233
18. 3,025 − 1,675 = 1,350
19. 3,672 + 8,074 = 11,746
20. 41,908 + 34,821 + 27,889 = 104,618

Total Problems: Total Correct: Score:

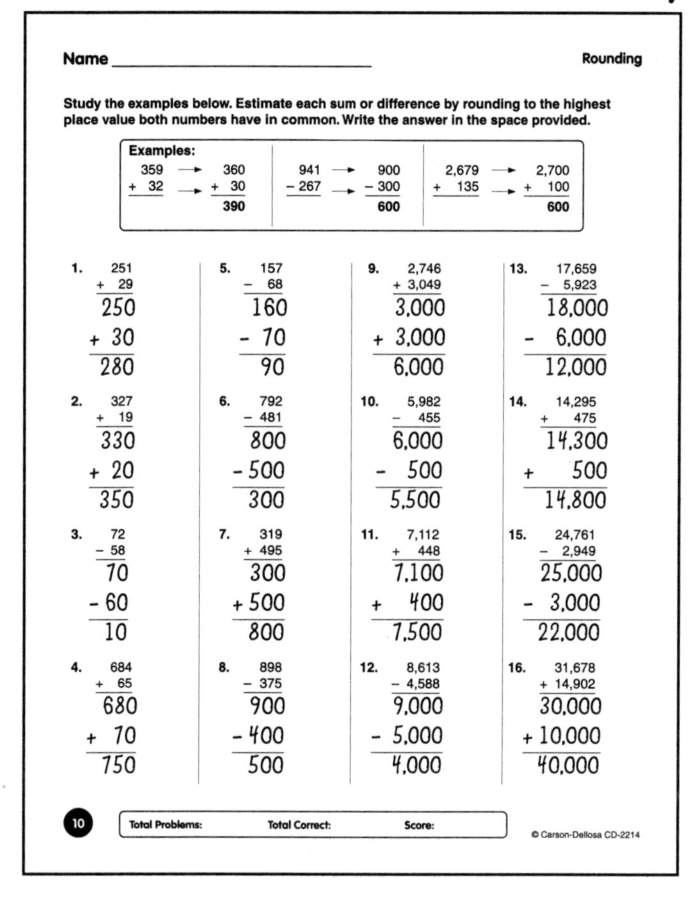

Name ____________________ **Rounding**

Study the examples below. Estimate each sum or difference by rounding to the highest place value both numbers have in common. Write the answer in the space provided.

Examples:

359	→ 360	941	→ 900	2,679	→ 2,700
+ 32	→ + 30	− 267	→ − 300	+ 135	→ + 100
	390		**600**		**600**

1. 251 + 29 → 250 + 30 = 280
2. 327 + 19 → 330 + 20 = 350
3. 72 − 58 → 70 − 60 = 10
4. 684 + 65 → 680 + 70 = 750
5. 157 − 68 → 160 − 70 = 90
6. 792 − 481 → 800 − 500 = 300
7. 319 + 495 → 300 + 500 = 800
8. 898 − 375 → 900 − 400 = 500
9. 2,746 + 3,049 → 3,000 + 3,000 = 6,000
10. 5,982 − 455 → 6,000 − 500 = 5,500
11. 7,112 + 448 → 7,100 + 400 = 7,500
12. 8,613 − 4,588 → 9,000 − 5,000 = 4,000
13. 17,659 − 5,923 → 18,000 − 6,000 = 12,000
14. 14,295 + 475 → 14,300 + 500 = 14,800
15. 24,761 − 2,949 → 25,000 − 3,000 = 22,000
16. 31,678 + 14,902 → 30,000 + 10,000 = 40,000

Total Problems: Total Correct: Score:

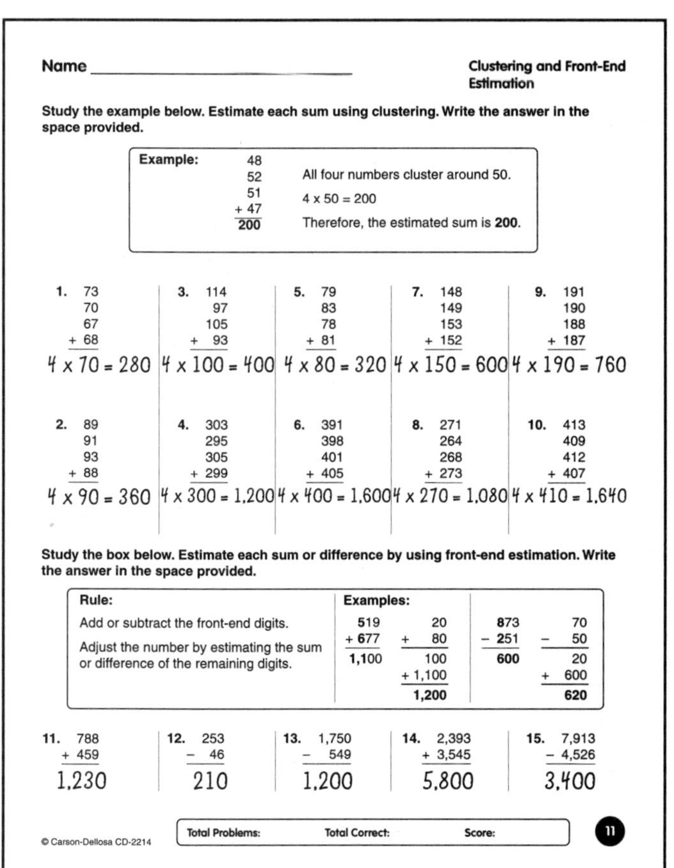

Name ____________________ **Clustering and Front-End Estimation**

Study the example below. Estimate each sum using clustering. Write the answer in the space provided.

Example: 48 + 52 + 51 + 47 = **200**

All four numbers cluster around 50.
4 x 50 = 200
Therefore, the estimated sum is **200**.

1. 73 + 70 + 67 + 68: 4 x 70 = 280
2. 89 + 91 + 93 + 88: 4 x 90 = 360
3. 114 + 97 + 105 + 93: 4 x 100 = 400
4. 303 + 295 + 305 + 299: 4 x 300 = 1,200
5. 79 + 83 + 78 + 81: 4 x 80 = 320
6. 391 + 398 + 401 + 405: 4 x 400 = 1,600
7. 148 + 149 + 153 + 152: 4 x 150 = 600
8. 271 + 264 + 268 + 273: 4 x 270 = 1,080
9. 191 + 190 + 188 + 187: 4 x 190 = 760
10. 413 + 409 + 412 + 407: 4 x 410 = 1,640

Study the box below. Estimate each sum or difference by using front-end estimation. Write the answer in the space provided.

Rule:
Add or subtract the front-end digits.
Adjust the number by estimating the sum or difference of the remaining digits.

Examples:
519 + 677 → **1,100**; 20 + 80 = 100; 100 + 1,100 = **1,200**
873 − 251 → **600**; 70 − 50 = 20; 20 + 600 = **620**

11. 788 + 459 = 1,230
12. 253 − 46 = 210
13. 1,750 − 549 = 1,200
14. 2,393 + 3,545 = 5,800
15. 7,913 − 4,526 = 3,400

Total Problems: Total Correct: Score:

Name ____________________ **Problem Solving with Whole Numbers**

Solve each word problem. Show your work and write the answer in the space provided.

1. Of the 1,329 students at Peaks Middle School, 483 voted in favor of having school uniforms. How many students voted against it?
   1,329 − 483 = 846 students

2. An amusement park had 39,345 people enter its gates on Saturday and 42,195 on Sunday. How many people walked through the gates in all?
   39,345 + 42,195 = 81,540 people

3. John purchased a new car at a cost of $29,199. If he had to pay $2,045 in tax, how much did he spend for the car?
   $29,199 + $2,045 = $31,244

4. During the food drive, Harris's school collected 2,451 cans of assorted food. His younger sister's school collected 1,973 cans of food. How many cans of food were collected in all?
   2,451 + 1,973 = 4,424 cans of food

5. Gary's company had sales of $178,354 during the month of October. He had expenses of $129,459. How much profit did he make after expenses?
   $178,354 − $129,459 = $48,895

6. Sean's salary is $57,000 per year. Iris makes $42,191 per year. How much more money does Sean make?
   $57,000 − $42,191 = $14,809

7. During his high school football career, David rushed for 3,841 yards. His younger brother Joe rushed for 4,095 yards. How many more yards did Joe rush?
   3,841 − 4,095 = 254 yards

8. Akeem put 11,951 miles on his new car during the first year. During the second year, he drove it 8,973 miles. How many miles did Akeem have on his car after the second year?
   11,451 + 8,973 = 20,924 miles

Total Problems: Total Correct: Score:

Name ____________________ **Multiplication**

Find the products. Write the answer in the space provided.

1. 24 x 9 = 216	6. 349 x 27 = 9,423	11. 672 x 315 = 211,680	16. 2,786 x 563 = 1,568,518
2. 73 x 8 = 584	7. 4,510 x 5 = 22,550	12. 820 x 141 = 115,620	17. 9,017 x 483 = 4,355,211
3. 42 x 6 = 252	8. 8,032 x 34 = 273,088	13. 6,509 x 446 = 2,903,014	18. 1,598 x 677 = 1,081,846
4. 583 x 5 = 2,915	9. 9,316 x 48 = 447,168	14. 5,592 x 700 = 3,914,400	19. 2,638 x 450 = 1,187,100
5. 415 x 72 = 29,880	10. 38 x 289 = 10,982	15. 642 x 526 = 337,692	20. 3,138 x 495 = 1,553,310

Total Problems: Total Correct: Score:

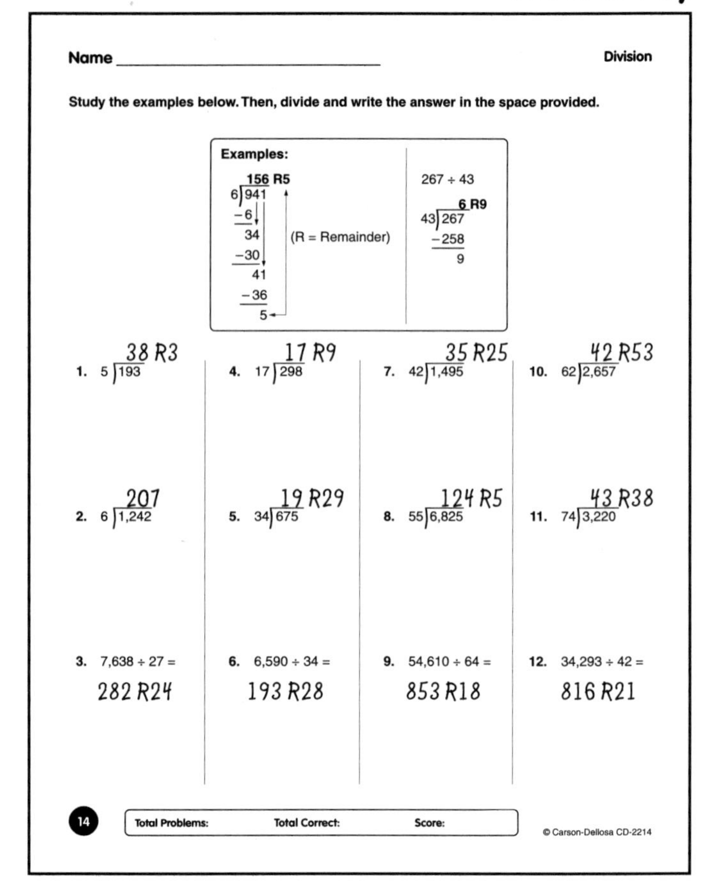

Name ____________________ **Division**

Study the examples below. Then, divide and write the answer in the space provided.

Examples:

6)941 = 156 R5: −6, 34, −30, 41, −36, 5 (R = Remainder)

267 ÷ 43: 43)267 = 6 R9: −258, 9

1. 5)193 = 38 R3	4. 17)298 = 17 R9	7. 42)1,495 = 35 R25	10. 62)2,657 = 42 R53
2. 6)1,242 = 207	5. 34)675 = 19 R29	8. 55)6,825 = 124 R5	11. 74)3,220 = 43 R38
3. 7,638 ÷ 27 = 282 R24	6. 6,590 ÷ 34 = 193 R28	9. 54,610 ÷ 64 = 853 R18	12. 34,293 ÷ 42 = 816 R21

Total Problems: Total Correct: Score:

Name ____________________ **Estimating Products**

Study the example below. Use rounding to estimate each product. Round to the highest place value but do not round single-digit numbers. Write the answer on the line provided.

Example:

275 x 9 → 300 x 9 3 x 9 = 27 plus 2 zeros, gives an estimate of **2,700**.

1. 28 x 7 = 210	10. 392 x 115 = 40,000
2. 21 x 17 = 400	11. 2,499 x 83 = 160,000
3. 185 x 2 = 400	12. 5,081 x 308 = 1,500,000
4. 218 x 41 = 8,000	13. 43 x 8 = 320
5. 1,321 x 63 = 60,000	14. 72 x 45 = 3,500
6. 2,812 x 718 = 2,100,00	15. 93 x 121 = 9,000
7. 15 x 19 = 400	16. 451 x 242 = 100,000
8. 39 x 41 = 1,600	17. 25 x 749 = 21,000
9. 231 x 7 = 1,400	18. 6,250 x 627 = 3,600,000

Total Problems: Total Correct: Score:

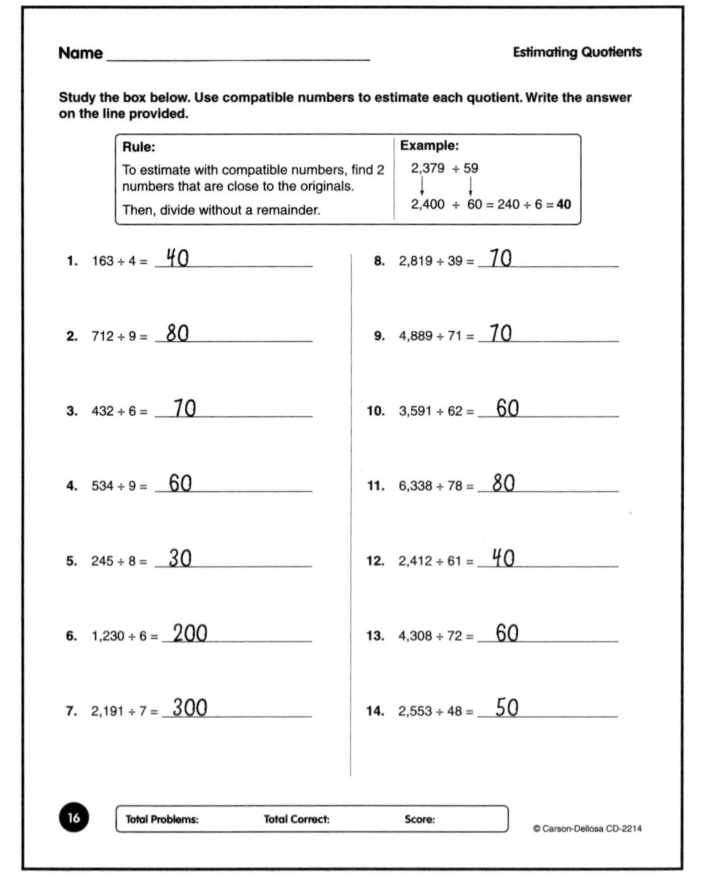

Name ____________________ **Estimating Quotients**

Study the box below. Use compatible numbers to estimate each quotient. Write the answer on the line provided.

Rule: To estimate with compatible numbers, find 2 numbers that are close to the originals. Then, divide without a remainder.

Example: 2,379 ÷ 59 → 2,400 ÷ 60 = 240 ÷ 6 = **40**

1. 163 ÷ 4 = 40	8. 2,819 ÷ 39 = 70
2. 712 ÷ 9 = 80	9. 4,889 ÷ 71 = 70
3. 432 ÷ 6 = 70	10. 3,591 ÷ 62 = 60
4. 534 ÷ 9 = 60	11. 6,338 ÷ 78 = 80
5. 245 ÷ 8 = 30	12. 2,412 ÷ 61 = 40
6. 1,230 ÷ 6 = 200	13. 4,308 ÷ 72 = 60
7. 2,191 ÷ 7 = 300	14. 2,553 ÷ 48 = 50

Total Problems: Total Correct: Score:

Name ______________________ **Order of Operations**

Study the box below. Evaluate each expression. Write the answer on the line provided.

Rule:	Examples:	
Order of Operations: 1. Work inside the parentheses. 2. Compute all exponents. 3. Multiply and divide from left to right. 4. Add and subtract from left to right.	$5^2 + 3 \cdot 2$ $25 + 3 \cdot 2$ $25 + 6 = $ **31**	$17 - 5 \cdot 3$ $17 - 15 = $ **2**

1. $2 + 5 \cdot 2^2 =$ 22
2. $(2 + 5) \cdot 2^2 =$ 28
3. $(3^3 - 5) \cdot 2 =$ 44
4. $3^3 - 5 \cdot 2 =$ 17
5. $3 \cdot (5^2 - 10) =$ 45
6. $3 \cdot 5^2 - 10 =$ 65
7. $25 \div (2 + 3)^2 =$ 1
8. $30 \div 2 + 2^2 =$ 19
9. $(8 - 2)^2 \cdot 2 =$ 72
10. $8 - 1 \cdot 4 + 3 =$ 7
11. $(8 - 1) \cdot 4 + 3 =$ 31
12. $(6^2 - 3) + 5 =$ 38
13. $4 + (2^2 - 3) + 5 =$ 10
14. $6^2 - 6 \div 3 =$ 34

Total Problems: Total Correct: Score:

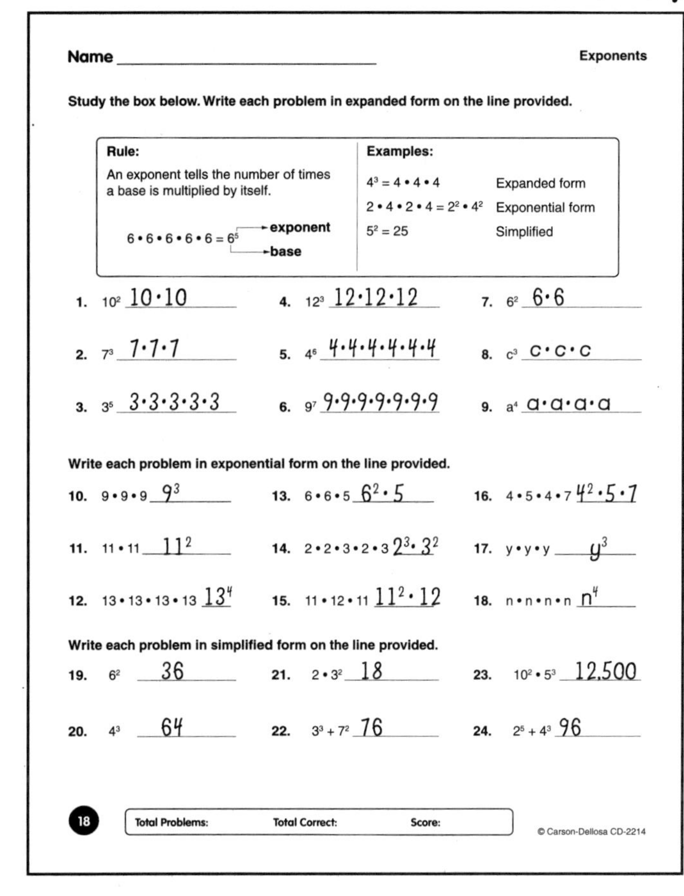

Name ______________________ **Exponents**

Study the box below. Write each problem in expanded form on the line provided.

Rule:	Examples:	
An exponent tells the number of times a base is multiplied by itself. $6 \cdot 6 \cdot 6 \cdot 6 \cdot 6 = 6^5$ (5 → exponent, 6 → base)	$4^3 = 4 \cdot 4 \cdot 4$ $2 \cdot 4 \cdot 2 \cdot 4 = 2^2 \cdot 4^2$ $5^2 = 25$	Expanded form Exponential form Simplified

1. 10^2 10·10
2. 7^3 7·7·7
3. 3^5 3·3·3·3·3
4. 12^3 12·12·12
5. 4^6 4·4·4·4·4·4
6. 9^7 9·9·9·9·9·9·9
7. 6^2 6·6
8. c^3 c·c·c
9. a^4 a·a·a·a

Write each problem in exponential form on the line provided.

10. $9 \cdot 9 \cdot 9$ 9^3
11. $11 \cdot 11$ 11^2
12. $13 \cdot 13 \cdot 13 \cdot 13$ 13^4
13. $6 \cdot 6 \cdot 5$ $6^2 \cdot 5$
14. $2 \cdot 2 \cdot 3 \cdot 2 \cdot 3$ $2^3 \cdot 3^2$
15. $11 \cdot 12 \cdot 11$ $11^2 \cdot 12$
16. $4 \cdot 5 \cdot 4 \cdot 7$ $4^2 \cdot 5 \cdot 7$
17. $y \cdot y \cdot y$ y^3
18. $n \cdot n \cdot n \cdot n$ n^4

Write each problem in simplified form on the line provided.

19. 6^2 36
20. 4^3 64
21. $2 \cdot 3^2$ 18
22. $3^3 + 7^2$ 76
23. $10^2 \cdot 5^3$ 12,500
24. $2^5 + 4^3$ 96

Total Problems: Total Correct: Score:

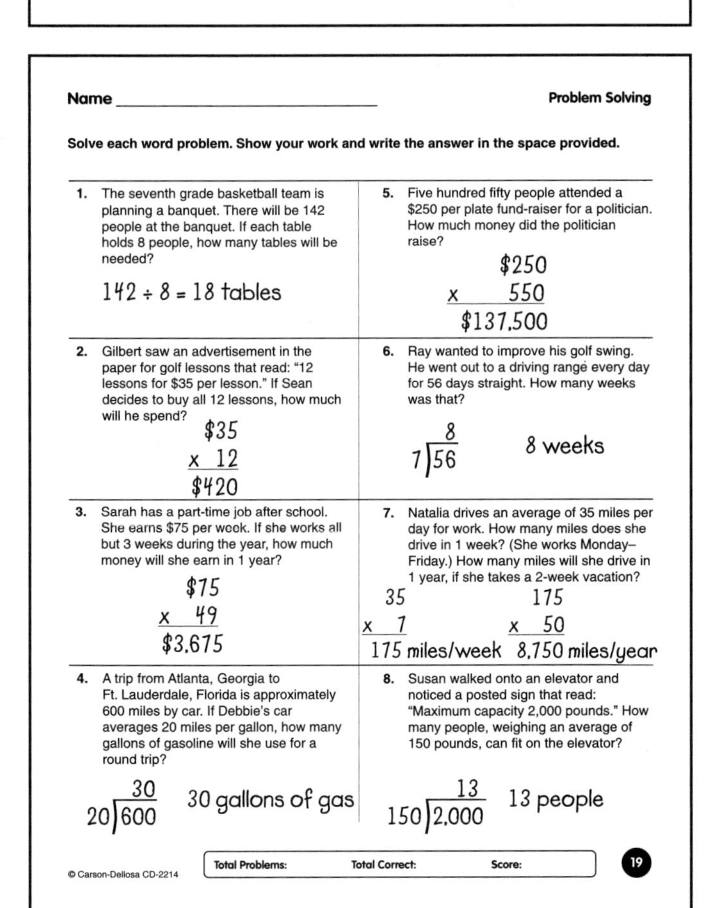

Name ______________________ **Problem Solving**

Solve each word problem. Show your work and write the answer in the space provided.

1. The seventh grade basketball team is planning a banquet. There will be 142 people at the banquet. If each table holds 8 people, how many tables will be needed?

   142 ÷ 8 = 18 tables

2. Gilbert saw an advertisement in the paper for golf lessons that read: "12 lessons for $35 per lesson." If Sean decides to buy all 12 lessons, how much will he spend?

   $35 × 12 = $420

3. Sarah has a part-time job after school. She earns $75 per week. If she works all but 3 weeks during the year, how much money will she earn in 1 year?

   $75 × 49 = $3,675

4. A trip from Atlanta, Georgia to Ft. Lauderdale, Florida is approximately 600 miles by car. If Debbie's car averages 20 miles per gallon, how many gallons of gasoline will she use for a round trip?

   600 ÷ 20 = 30 — 30 gallons of gas

5. Five hundred fifty people attended a $250 per plate fund-raiser for a politician. How much money did the politician raise?

   $250 × 550 = $137,500

6. Ray wanted to improve his golf swing. He went out to a driving range every day for 56 days straight. How many weeks was that?

   56 ÷ 7 = 8 — 8 weeks

7. Natalia drives an average of 35 miles per day for work. How many miles does she drive in 1 week? (She works Monday–Friday.) How many miles will she drive in 1 year, if she takes a 2-week vacation?

   35 × 7 = 175 miles/week; 175 × 50 = 8,750 miles/year

8. Susan walked onto an elevator and noticed a posted sign that read: "Maximum capacity 2,000 pounds." How many people, weighing an average of 150 pounds, can fit on the elevator?

   2,000 ÷ 150 = 13 — 13 people

Total Problems: Total Correct: Score:

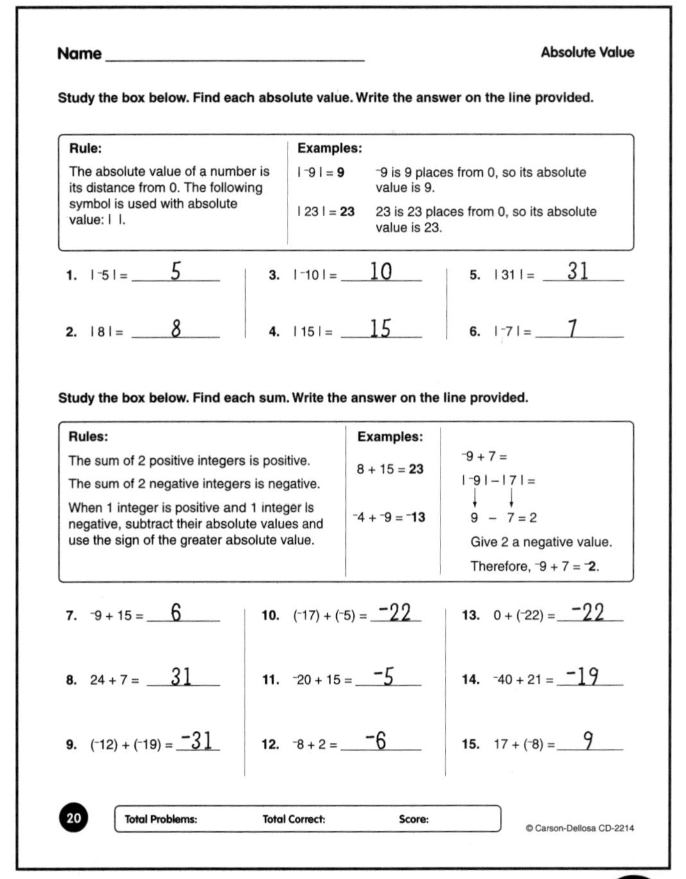

Name ______________________ **Absolute Value**

Study the box below. Find each absolute value. Write the answer on the line provided.

Rule:	Examples:	
The absolute value of a number is its distance from 0. The following symbol is used with absolute value: | |.	|⁻9| = **9** |23| = **23**	⁻9 is 9 places from 0, so its absolute value is 9. 23 is 23 places from 0, so its absolute value is 23.

1. \|⁻5\| = 5
2. \|8\| = 8
3. \|⁻10\| = 10
4. \|15\| = 15
5. \|31\| = 31
6. \|⁻7\| = 7

Study the box below. Find each sum. Write the answer on the line provided.

Rules:	Examples:	
The sum of 2 positive integers is positive. The sum of 2 negative integers is negative. When 1 integer is positive and 1 integer is negative, subtract their absolute values and use the sign of the greater absolute value.	8 + 15 = **23** ⁻4 + ⁻9 = **⁻13**	⁻9 + 7 = |⁻9| – |7| = 9 – 7 = 2 Give 2 a negative value. Therefore, ⁻9 + 7 = **⁻2**.

7. ⁻9 + 15 = 6
8. 24 + 7 = 31
9. (⁻12) + (⁻19) = ⁻31
10. (⁻17) + (⁻5) = ⁻22
11. ⁻20 + 15 = ⁻5
12. ⁻8 + 2 = ⁻6
13. 0 + (⁻22) = ⁻22
14. ⁻40 + 21 = ⁻19
15. 17 + (⁻8) = 9

Total Problems: Total Correct: Score:

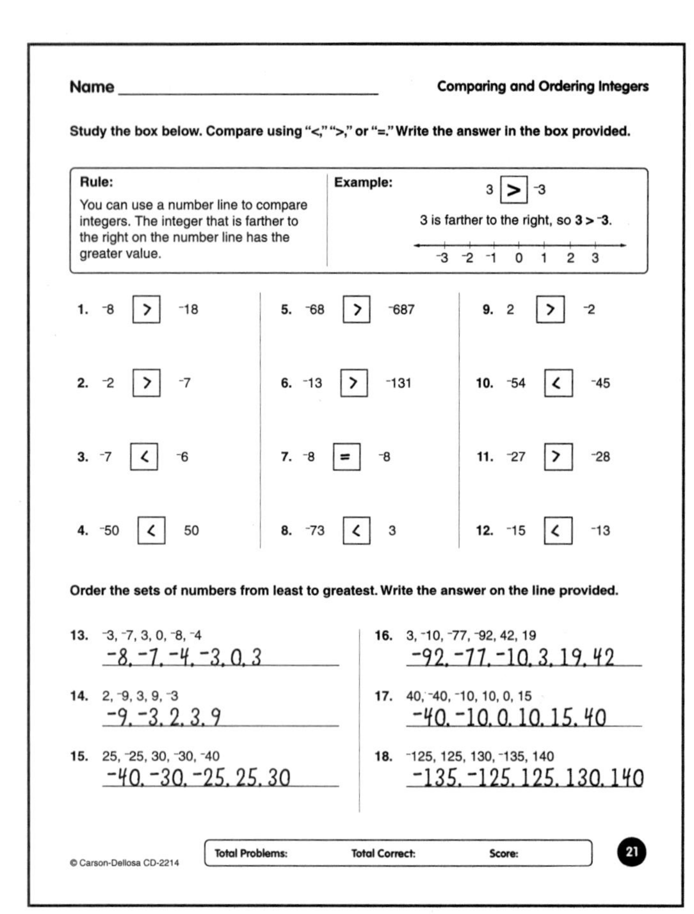

Name ______________________ **Comparing and Ordering Integers**

Study the box below. Compare using "<," ">," or "=." Write the answer in the box provided.

Rule:	Example:
You can use a number line to compare integers. The integer that is farther to the right on the number line has the greater value.	3 [>] -3 3 is farther to the right, so **3 > -3**. -3 -2 -1 0 1 2 3

1. -8 [>] -18	5. -68 [>] -687	9. 2 [>] -2
2. -2 [>] -7	6. -13 [>] -131	10. -54 [<] -45
3. -7 [<] -6	7. -8 [=] -8	11. -27 [>] -28
4. -50 [<] 50	8. -73 [<] 3	12. -15 [<] -13

Order the sets of numbers from least to greatest. Write the answer on the line provided.

13. -3, -7, 3, 0, -8, -4
-8, -7, -4, -3, 0, 3

14. 2, -9, 3, 9, -3
-9, -3, 2, 3, 9

15. 25, -25, 30, -30, -40
-40, -30, -25, 25, 30

16. 3, -10, -77, -92, 42, 19
-92, -77, -10, 3, 19, 42

17. 40, -40, -10, 10, 0, 15
-40, -10, 0, 10, 15, 40

18. -125, 125, 130, -135, 140
-135, -125, 125, 130, 140

© Carson-Dellosa CD-2214 Total Problems: Total Correct: Score: 21

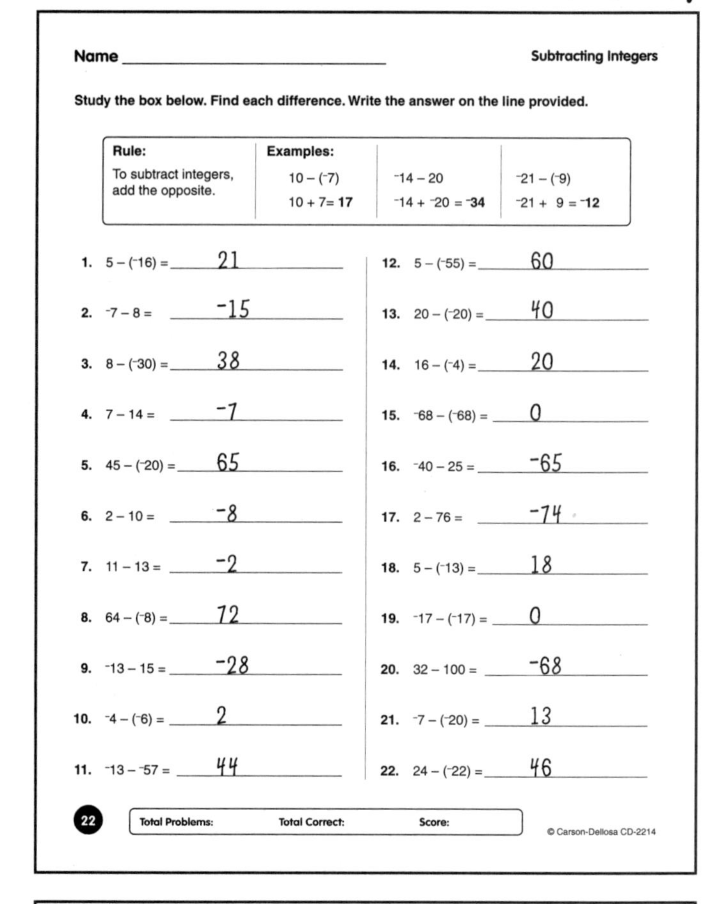

Name ______________________ **Subtracting Integers**

Study the box below. Find each difference. Write the answer on the line provided.

Rule:	Examples:		
To subtract integers, add the opposite.	10 – (-7) 10 + 7 = **17**	-14 – 20 -14 + -20 = **-34**	-21 – (-9) -21 + 9 = **-12**

1. 5 – (-16) = 21	12. 5 – (-55) = 60
2. -7 – 8 = -15	13. 20 – (-20) = 40
3. 8 – (-30) = 38	14. 16 – (-4) = 20
4. 7 – 14 = -7	15. -68 – (-68) = 0
5. 45 – (-20) = 65	16. -40 – 25 = -65
6. 2 – 10 = -8	17. 2 – 76 = -74
7. 11 – 13 = -2	18. 5 – (-13) = 18
8. 64 – (-8) = 72	19. -17 – (-17) = 0
9. -13 – 15 = -28	20. 32 – 100 = -68
10. -4 – (-6) = 2	21. -7 – (-20) = 13
11. -13 – -57 = 44	22. 24 – (-22) = 46

22 Total Problems: Total Correct: Score: © Carson-Dellosa CD-2214

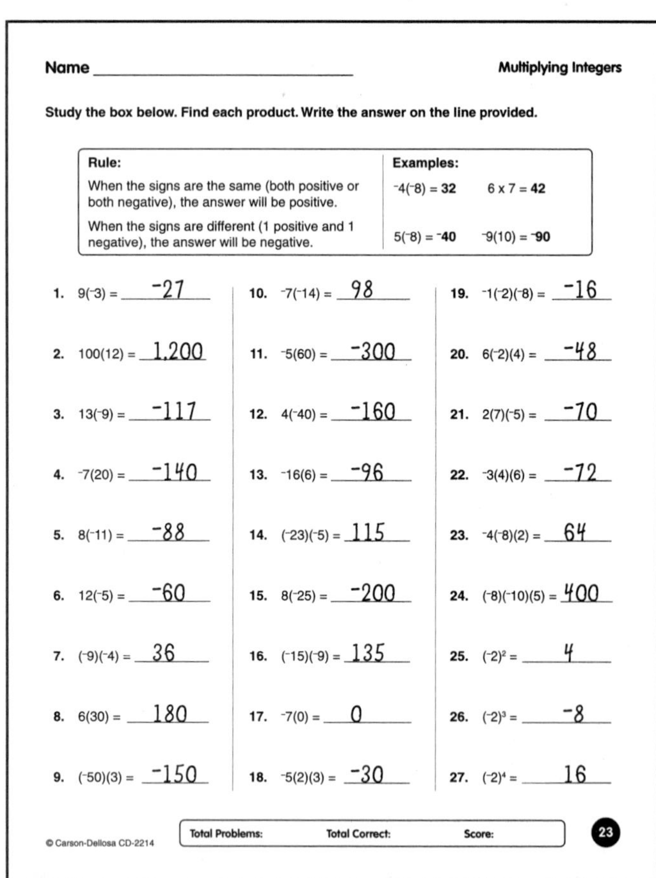

Name ______________________ **Multiplying Integers**

Study the box below. Find each product. Write the answer on the line provided.

Rule:	Examples:	
When the signs are the same (both positive or both negative), the answer will be positive.	-4(-8) = **32**	6 x 7 = **42**
When the signs are different (1 positive and 1 negative), the answer will be negative.	5(-8) = **-40**	-9(10) = **-90**

1. 9(-3) = -27	10. -7(-14) = 98	19. -1(-2)(-8) = -16
2. 100(12) = 1,200	11. -5(60) = -300	20. 6(-2)(4) = -48
3. 13(-9) = -117	12. 4(-40) = -160	21. 2(7)(-5) = -70
4. -7(20) = -140	13. -16(6) = -96	22. -3(4)(6) = -72
5. 8(-11) = -88	14. (-23)(-5) = 115	23. -4(-8)(2) = 64
6. 12(-5) = -60	15. 8(-25) = -200	24. (-8)(-10)(5) = 400
7. (-9)(-4) = 36	16. (-15)(-9) = 135	25. $(-2)^2$ = 4
8. 6(30) = 180	17. -7(0) = 0	26. $(-2)^3$ = -8
9. (-50)(3) = -150	18. -5(2)(3) = -30	27. $(-2)^4$ = 16

© Carson-Dellosa CD-2214 Total Problems: Total Correct: Score: 23

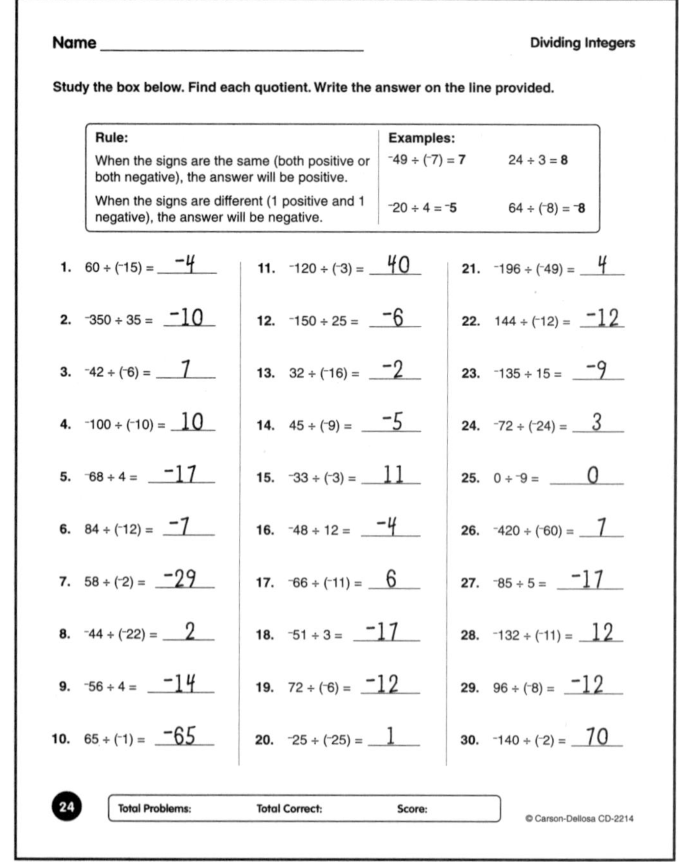

Name ______________________ **Dividing Integers**

Study the box below. Find each quotient. Write the answer on the line provided.

Rule:	Examples:	
When the signs are the same (both positive or both negative), the answer will be positive.	-49 ÷ (-7) = **7**	24 ÷ 3 = **8**
When the signs are different (1 positive and 1 negative), the answer will be negative.	-20 ÷ 4 = **-5**	64 ÷ (-8) = **-8**

1. 60 ÷ (-15) = -4	11. -120 ÷ (-3) = 40	21. -196 ÷ (-49) = 4
2. -350 ÷ 35 = -10	12. -150 ÷ 25 = -6	22. 144 ÷ (-12) = -12
3. -42 ÷ (-6) = 7	13. 32 ÷ (-16) = -2	23. -135 ÷ 15 = -9
4. -100 ÷ (-10) = 10	14. 45 ÷ (-9) = -5	24. -72 ÷ (-24) = 3
5. -68 ÷ 4 = -17	15. -33 ÷ (-3) = 11	25. 0 ÷ -9 = 0
6. 84 ÷ (-12) = -7	16. -48 ÷ 12 = -4	26. -420 ÷ (-60) = 7
7. 58 ÷ (-2) = -29	17. -66 ÷ (-11) = 6	27. -85 ÷ 5 = -17
8. -44 ÷ (-22) = 2	18. -51 ÷ 3 = -17	28. -132 ÷ (-11) = 12
9. -56 ÷ 4 = -14	19. 72 ÷ (-6) = -12	29. 96 ÷ (-8) = -12
10. 65 ÷ (-1) = -65	20. -25 ÷ (-25) = 1	30. -140 ÷ (-2) = 70

24 Total Problems: Total Correct: Score: © Carson-Dellosa CD-2214

Name ______________________________ **Problem Solving with Integers**

Solve each word problem. Show your work and write the answer in the space provided.

1. Raoul had a balance of \$532 in his checking account. He wrote a check for \$73. Write an addition sentence which represents this situation. How much money does he have left in his account?

   \$532 + (−\$73) = \$459

2. During a golf tournament, Dave scored 2 under par (−2) each day of the 4 days of the tournament. Express as an integer how many strokes he finished under par.

   $4 \times {}^{-}2 = {}^{-}8$

3. Find each product.

   A. $(^{-}4)^2$ $^{-}4 \cdot {}^{-}4 = 16$

   B. $(^{-}4)^3$ $^{-}4 \cdot {}^{-}4 \cdot {}^{-}4 = {}^{-}64$

   C. $(^{-}5)^2$ $^{-}5 \cdot {}^{-}5 = 25$

   D. $(^{-}5)^3$ $^{-}5 \cdot {}^{-}5 \cdot {}^{-}5 = {}^{-}125$

4. Find the product of any 2 negative factors, any 3 negative factors, any 4 factors, and any 5 negative factors. Look for a pattern in each product. Write a rule for multiplying more than 2 negative factors.

   Answers will vary.

5. Ethel invested in a stock that was recommended to her by her broker. During the first 5 days of owning the stock, it dropped 10 points (−10). What was the average change per day?

   $^{-}10 \div 5 = {}^{-}2$ −2 per day

6. What is the sum of an integer and its opposite? Give examples to support your answer.

   0

   Examples will vary.

7. Write the next 2 terms in each sequence.

   A. 3, 2, 1, 0, −1, −2, −3

   B. 5, 4, 2, −1, −5, −10

   C. −16, −14, −12, −10, −8, −6

8. When Charlotte went to bed, it was 9°F. When she woke up in the morning, the temperature had dropped to −2°F. How many degrees did the temperature drop?

   $x = 9° - (^{-}2°)$

   $x = 9° + 2°$

   $x = 11°F$

Total Problems: Total Correct: Score:

Name ______________________________ **Equivalent Fractions**

Study the box below. Find the missing numerator or denominator to make equivalent fractions. Write the answer in the space provided.

Rule:	Example:
Multiply (or divide) the numerator and denominator by the same number to make equivalent fractions.	$\frac{4}{5} = \frac{\ }{15}$ $\frac{4 \times 3}{5 \times 3} = \frac{12}{15}$

1. $\frac{3}{6} = \frac{6}{12}$
2. $\frac{5}{11} = \frac{20}{44}$
3. $\frac{3}{10} = \frac{18}{60}$
4. $\frac{28}{32} = \frac{7}{8}$
5. $\frac{7}{8} = \frac{49}{56}$
6. $\frac{8}{15} = \frac{64}{120}$
7. $\frac{24}{32} = \frac{6}{8}$
8. $\frac{15}{60} = \frac{1}{4}$
9. $\frac{10}{7} = \frac{70}{49}$
10. $\frac{2}{3} = \frac{6}{9} = \frac{8}{12}$
11. $\frac{7}{8} = \frac{21}{24}$
12. $\frac{6}{7} = \frac{30}{35}$
13. $\frac{24}{30} = \frac{4}{5}$
14. $\frac{17}{34} = \frac{2}{4}$
15. $\frac{6}{13} = \frac{12}{26}$
16. $\frac{3}{8} = \frac{15}{40}$
17. $\frac{9}{12} = \frac{27}{36}$
18. $\frac{9}{9} = \frac{25}{25}$
19. $\frac{12}{4} = \frac{24}{8} = \frac{120}{40}$
20. $\frac{3}{4} = \frac{6}{8} = \frac{24}{32}$
21. $\frac{4}{9} = \frac{16}{36}$
22. $\frac{8}{9} = \frac{40}{45}$
23. $\frac{5}{12} = \frac{20}{48}$
24. $\frac{27}{45} = \frac{3}{5}$
25. $\frac{9}{11} = \frac{54}{66}$
26. $\frac{18}{48} = \frac{3}{8}$
27. $\frac{11}{15} = \frac{44}{60}$
28. $\frac{4}{16} = \frac{1}{4}$
29. $\frac{7}{9} = \frac{42}{54}$
30. $\frac{9}{10} = \frac{27}{30} = \frac{36}{40}$

Total Problems: Total Correct: Score:

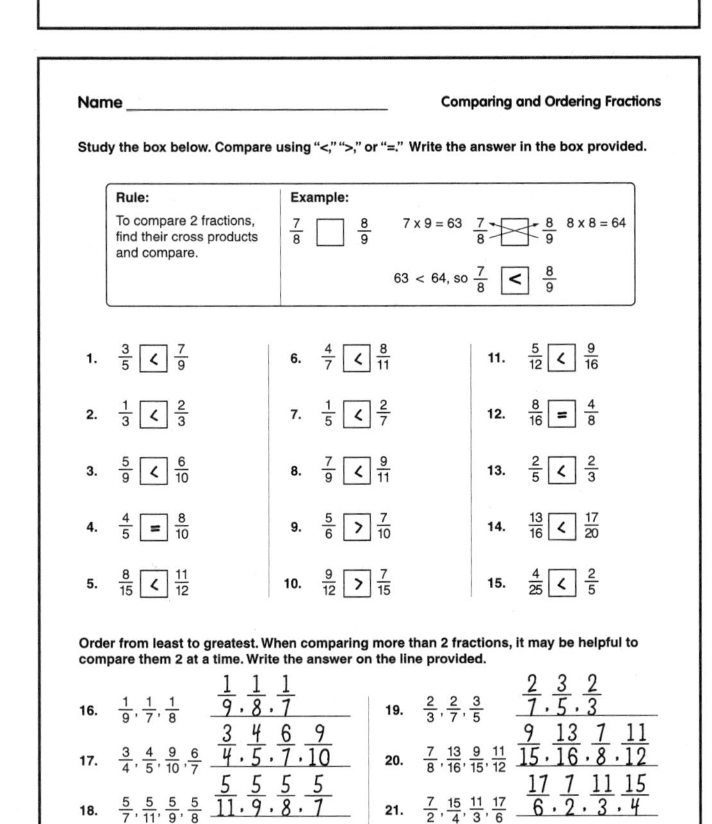

Name ______________________________ **Comparing and Ordering Fractions**

Study the box below. Compare using "<," ">," or "=." Write the answer in the box provided.

Rule:	Example:
To compare 2 fractions, find their cross products and compare.	$\frac{7}{8} \square \frac{8}{9}$ 7 x 9 = 63 8 x 8 = 64 63 < 64, so $\frac{7}{8} < \frac{8}{9}$

1. $\frac{3}{5} < \frac{7}{9}$
2. $\frac{1}{3} < \frac{2}{3}$
3. $\frac{5}{9} < \frac{6}{10}$
4. $\frac{4}{5} = \frac{8}{10}$
5. $\frac{8}{15} < \frac{11}{12}$
6. $\frac{4}{7} < \frac{8}{11}$
7. $\frac{1}{5} < \frac{2}{7}$
8. $\frac{7}{9} < \frac{9}{11}$
9. $\frac{5}{6} > \frac{7}{10}$
10. $\frac{9}{12} > \frac{7}{15}$
11. $\frac{5}{12} < \frac{9}{16}$
12. $\frac{8}{16} = \frac{4}{8}$
13. $\frac{2}{5} < \frac{2}{3}$
14. $\frac{13}{16} < \frac{17}{20}$
15. $\frac{4}{25} < \frac{2}{5}$

Order from least to greatest. When comparing more than 2 fractions, it may be helpful to compare them 2 at a time. Write the answer on the line provided.

16. $\frac{1}{9}, \frac{1}{7}, \frac{1}{8}$ $\frac{1}{9}, \frac{1}{8}, \frac{1}{7}$
17. $\frac{3}{4}, \frac{4}{5}, \frac{9}{10}, \frac{6}{7}$ $\frac{3}{4}, \frac{4}{5}, \frac{6}{7}, \frac{9}{10}$
18. $\frac{5}{7}, \frac{5}{11}, \frac{5}{9}, \frac{5}{8}$ $\frac{5}{11}, \frac{5}{9}, \frac{5}{8}, \frac{5}{7}$
19. $\frac{2}{3}, \frac{2}{7}, \frac{3}{5}$ $\frac{2}{7}, \frac{3}{5}, \frac{2}{3}$
20. $\frac{7}{8}, \frac{13}{16}, \frac{9}{15}, \frac{11}{12}$ $\frac{9}{15}, \frac{13}{16}, \frac{7}{8}, \frac{11}{12}$
21. $\frac{7}{2}, \frac{15}{4}, \frac{11}{3}, \frac{17}{6}$ $\frac{17}{6}, \frac{7}{2}, \frac{11}{3}, \frac{15}{4}$

Total Problems: Total Correct: Score:

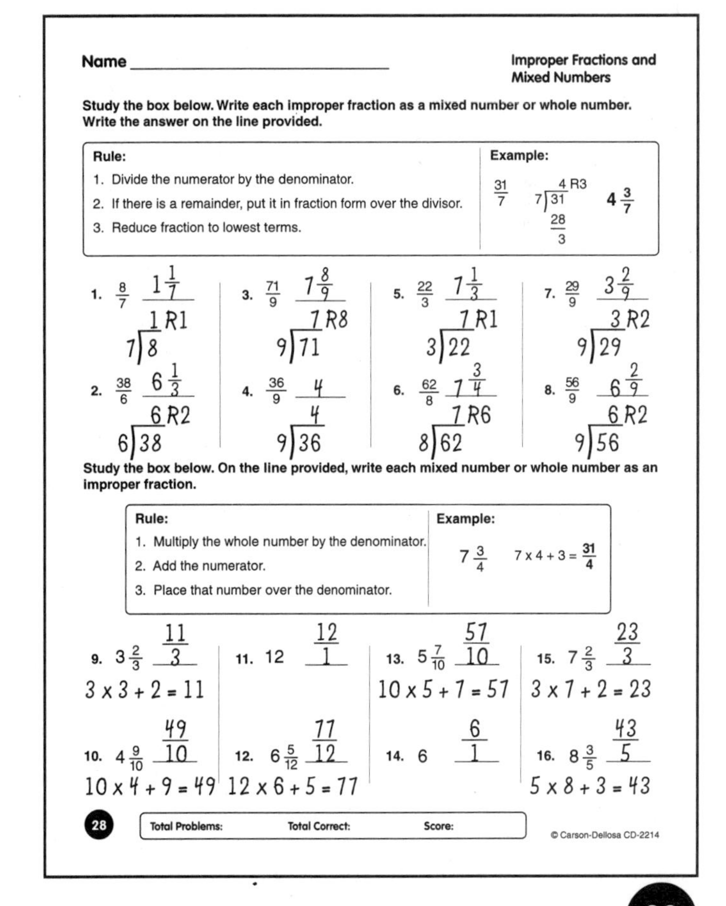

Name ______________________________ **Improper Fractions and Mixed Numbers**

Study the box below. Write each improper fraction as a mixed number or whole number. Write the answer on the line provided.

Rule:	Example:
1. Divide the numerator by the denominator. 2. If there is a remainder, put it in fraction form over the divisor. 3. Reduce fraction to lowest terms.	$\frac{31}{7}$ 31 ÷ 7 = 4 R3 (28, remainder 3) $4\frac{3}{7}$

1. $\frac{8}{7} = 1\frac{1}{7}$ (8 ÷ 7 = 1 R1)
2. $\frac{38}{6} = 6\frac{1}{3}$ (38 ÷ 6 = 6 R2)
3. $\frac{71}{9} = 7\frac{8}{9}$ (71 ÷ 9 = 7 R8)
4. $\frac{36}{9} = 4$ (36 ÷ 9 = 4)
5. $\frac{22}{3} = 7\frac{1}{3}$ (22 ÷ 3 = 7 R1)
6. $\frac{62}{8} = 7\frac{3}{4}$ (62 ÷ 8 = 7 R6)
7. $\frac{29}{9} = 3\frac{2}{9}$ (29 ÷ 9 = 3 R2)
8. $\frac{56}{9} = 6\frac{2}{9}$ (56 ÷ 9 = 6 R2)

Study the box below. On the line provided, write each mixed number or whole number as an improper fraction.

Rule:	Example:
1. Multiply the whole number by the denominator. 2. Add the numerator. 3. Place that number over the denominator.	$7\frac{3}{4}$ $7 \times 4 + 3 = \frac{31}{4}$

9. $3\frac{2}{3} = \frac{11}{3}$ 3 x 3 + 2 = 11
10. $4\frac{9}{10} = \frac{49}{10}$ 10 x 4 + 9 = 49
11. $12 = \frac{12}{1}$
12. $6\frac{5}{12} = \frac{77}{12}$ 12 x 6 + 5 = 77
13. $5\frac{7}{10} = \frac{57}{10}$ 10 x 5 + 7 = 57
14. $6 = \frac{6}{1}$
15. $7\frac{2}{3} = \frac{23}{3}$ 3 x 7 + 2 = 23
16. $8\frac{3}{5} = \frac{43}{5}$ 5 x 8 + 3 = 43

Total Problems: Total Correct: Score:

Name ______________________ **Estimating Fractions**

Study the examples below. Round each fraction to 0, $\frac{1}{2}$, or 1. Write the answer on the line provided.

Examples: $\frac{1}{8} \to 0$	$\frac{3}{5} \to \frac{1}{2}$	$\frac{11}{12} \to 1$
The numerator is much smaller than the denominator.	The numerator is about half of the denominator.	The numerator and denominator are close in value.

1. $\frac{3}{4}$ 1
2. $\frac{1}{5}$ 0
3. $\frac{5}{8}$ $\frac{1}{2}$
4. $\frac{2}{9}$ 0
5. $\frac{3}{10}$ $\frac{1}{2}$
6. $\frac{7}{12}$ $\frac{1}{2}$
7. $\frac{5}{6}$ 1
8. $\frac{1}{4}$ $\frac{1}{2}$
9. $\frac{1}{10}$ 0
10. $\frac{7}{8}$ 1
11. $\frac{1}{7}$ 0
12. $\frac{4}{9}$ $\frac{1}{2}$

Study the box below. Estimate each sum or difference. Write the answer on the line provided.

Rule:	Example:
Round each fraction to 0, $\frac{1}{2}$, or 1. Then, add.	$\frac{3}{8} + \frac{5}{6} =$ ↓ ↓ $\frac{1}{2} + 1 = 1\frac{1}{2}$

13. $\frac{4}{5} - \frac{1}{2} = \frac{1}{2}$
14. $\frac{3}{4} + \frac{1}{3} = 1\frac{1}{2}$
15. $\frac{1}{2} + \frac{7}{8} = 1\frac{1}{2}$
16. $\frac{7}{8} - \frac{1}{3} = \frac{1}{2}$
17. $\frac{3}{8} - \frac{1}{10} = \frac{1}{2}$
18. $\frac{5}{8} - \frac{1}{12} = \frac{1}{2}$

Total Problems: Total Correct: Score:

Name ______________________ **Adding and Subtracting Fractions**

Study the box below. Find each sum or difference. Reduce the answer to lowest terms. Write the answer in the space provided.

Rule:	Examples:
1. Change any mixed numbers to improper fractions. 2. Find the Least Common Denominator (LCD) and rewrite the fractions. 3. Add or subtract. 4. Reduce if necessary.	$5\frac{5}{6} = \frac{35}{6} = \frac{35}{6}$ $-3\frac{2}{3} = \frac{11}{3} \times \frac{2}{2} = -\frac{22}{6}$ $\frac{13}{6} = 2\frac{1}{6}$

1. $4\frac{2}{9} + 5\frac{5}{9} = \frac{38}{9} + \frac{50}{9} = \frac{88}{9} = 9\frac{7}{9}$
2. $4\frac{5}{12} + 3\frac{8}{9} = \frac{159}{36} + \frac{140}{36} = \frac{299}{36} = 8\frac{11}{36}$
3. $5\frac{1}{5} + 5\frac{1}{3} + 3\frac{1}{15} =$ $\frac{78}{15} + \frac{80}{15} + \frac{80}{15} = \frac{204}{15}$ $\frac{204}{15} = 13\frac{3}{5}$
4. $7\frac{3}{8} + 5\frac{2}{5} = \frac{295}{40} + \frac{216}{40} = \frac{511}{40} = 12\frac{31}{40}$
5. $5\frac{1}{6} - 3\frac{2}{3} = \frac{31}{6} - \frac{22}{6} = \frac{9}{6} = 1\frac{1}{2}$
6. $12 - 5\frac{4}{9} =$ $\frac{108}{9} - \frac{49}{9} = \frac{59}{9}$ $\frac{59}{9} = 6\frac{5}{9}$
7. $12\frac{9}{10} - 10\frac{1}{5} = \frac{129}{10} - \frac{102}{10} = \frac{27}{10} = 2\frac{7}{10}$
8. $4\frac{3}{10} + 2\frac{1}{12} = \frac{258}{60} + \frac{125}{60} = \frac{383}{60} = 6\frac{23}{60}$
9. $5\frac{1}{4} + 6\frac{3}{10} + 3\frac{5}{8} =$ $\frac{210}{40} + \frac{252}{40} + \frac{145}{40} = \frac{607}{40}$ $\frac{607}{40} = 15\frac{7}{40}$

Total Problems: Total Correct: Score:

Name ______________________ **Multiplying Fractions**

Study the box below. Find each product and reduce to lowest terms. Write the answer in the space provided.

Rule:	Example:
1. Change each mixed number to an improper fraction. 2. Multiply the numerators. 3. Multiply the denominators. 4. Reduce if possible.	$4\frac{2}{3} \times 5 =$ $\frac{14}{3} \times \frac{5}{1} = \frac{70}{3} = 23\frac{1}{3}$

1. $\frac{2}{5} \times \frac{3}{4} = \frac{6}{20} = \frac{3}{10}$
2. $3\frac{2}{5} \times 4\frac{2}{3} =$ $\frac{17}{5} \times \frac{14}{3} = 15\frac{13}{15}$
3. $1\frac{2}{3} \times 3\frac{3}{4} =$ $\frac{5}{3} \times \frac{15}{4} = 6\frac{1}{4}$
4. $4\frac{3}{8} \times 2\frac{1}{3} =$ $\frac{35}{8} \times \frac{7}{3} = 10\frac{5}{24}$
5. $6\frac{4}{9} \times 2\frac{2}{3} =$ $\frac{58}{9} \times \frac{8}{3} = 17\frac{5}{27}$
6. $4\frac{1}{6} \times 3\frac{2}{3} =$ $\frac{25}{6} \times \frac{11}{3} = 15\frac{5}{18}$
7. $5\frac{2}{5} \times 4\frac{1}{3} =$ $\frac{27}{5} \times \frac{13}{3} = 23\frac{2}{5}$
8. $5 \times 3\frac{2}{5} =$ $\frac{5}{1} \times \frac{17}{5} = 17$
9. $5\frac{3}{4} \times 2\frac{1}{2} =$ $\frac{23}{4} \times \frac{5}{2} = 14\frac{3}{8}$
10. $1\frac{4}{5} \times 2\frac{3}{10} =$ $\frac{9}{5} \times \frac{23}{10} = 4\frac{7}{50}$
11. $5\frac{3}{8} \times 4\frac{1}{3} =$ $\frac{43}{8} \times \frac{13}{3} = 23\frac{7}{24}$
12. $3\frac{1}{5} \times 5\frac{5}{8} =$ $\frac{16}{5} \times \frac{45}{8} = 18$
13. $8\frac{1}{4} \times 1\frac{1}{6} =$ $\frac{33}{4} \times \frac{7}{6} = 9\frac{5}{8}$
14. $9 \times 4\frac{1}{4} =$ $\frac{9}{1} \times \frac{17}{4} = 38\frac{1}{4}$
15. $3\frac{1}{4} \times 4 =$ $\frac{13}{4} \times \frac{4}{1} = 13$

Total Problems: Total Correct: Score:

Name ______________________ **Dividing Fractions**

Study the box below. Find each quotient and reduce to lowest terms. Write the answer in the space provided.

Rule:	Example:
1. Write the mixed numbers (or whole numbers) as improper fractions. 2. To divide fractions, flip the second fraction and change the division sign to multiplication. 3. Multiply, then reduce. Remember: A whole number can be written as a fraction by placing it over 1.	$9 \div 5\frac{2}{3}$ $\frac{9}{1} \div \frac{17}{3}$ $\frac{9}{1} \times \frac{3}{17} = \frac{27}{17} = 1\frac{27}{17}$

1. $3\frac{1}{6} \div \frac{2}{3} =$ $\frac{19}{6} \times \frac{3}{2} = 4\frac{3}{4}$
2. $3\frac{1}{2} \div 4\frac{3}{8} =$ $\frac{7}{2} \times \frac{8}{35} = \frac{4}{5}$
3. $6\frac{1}{8} \div 2\frac{1}{4} =$ $\frac{49}{8} \times \frac{4}{9} = 2\frac{13}{18}$
4. $7\frac{3}{10} \div 5\frac{2}{5} =$ $\frac{73}{10} \times \frac{5}{27} = 1\frac{19}{54}$
5. $12 \div 5\frac{5}{6} =$ $\frac{12}{1} \times \frac{6}{35} = 2\frac{2}{35}$
6. $7\frac{2}{9} \div \frac{5}{9} =$ $\frac{65}{9} \times \frac{9}{5} = 13$
7. $9\frac{1}{3} \div 2\frac{7}{9} =$ $\frac{28}{3} \times \frac{9}{25} = 3\frac{9}{25}$
8. $9\frac{6}{7} \div 2\frac{2}{7} =$ $\frac{69}{7} \times \frac{7}{16} = 4\frac{5}{16}$
9. $\frac{6}{11} \div 3\frac{4}{11} =$ $\frac{6}{11} \times \frac{11}{37} = \frac{6}{37}$
10. $7\frac{8}{9} \div 3\frac{2}{3} =$ $\frac{71}{9} \times \frac{3}{11} = 2\frac{5}{33}$
11. $21 \div 5\frac{1}{2} =$ $\frac{21}{1} \times \frac{2}{11} = 3\frac{9}{11}$
12. $18 \div 9\frac{1}{4} =$ $\frac{18}{1} \times \frac{4}{37} = 1\frac{35}{37}$
13. $4\frac{4}{5} \div \frac{3}{25} =$ $\frac{24}{5} \times \frac{25}{3} = 40$
14. $2\frac{7}{9} \div 11\frac{2}{3} =$ $\frac{25}{9} \times \frac{3}{35} = \frac{5}{21}$
15. $6\frac{8}{11} \div \frac{1}{11} =$ $\frac{74}{11} \times \frac{11}{1} = 74$

Total Problems: Total Correct: Score:

Name ____________________ **Problem Solving with Fractions**

Solve each word problem. Show your work and write the answer in the space provided.

1. Steven works for a land developer. The developer has purchased 12 acres of land on which to build houses. If each house is to lie on a $\frac{3}{4}$-acre lot, how many houses will the developer be able to build?

   $12 \div \frac{3}{4} = \frac{12}{1} \times \frac{3}{4} = 16$ houses

2. Maria is preparing cookies for her class. Her recipe calls for $1\frac{1}{2}$ cups of flour and $\frac{2}{3}$ cup of sugar per 24 cookies. If she needs to make 8 dozen cookies, how many cups of flour will she need? How much sugar?

   $4 \times \frac{3}{2} = 6$ cups of flour

   $4 \times \frac{2}{3} = 2\frac{2}{3}$ cups of sugar

3. Dave spent $6\frac{1}{2}$ hours at the park on Saturday and $7\frac{3}{4}$ hours on Sunday. How many hours did he spend in all?

   $\frac{26}{4} + \frac{31}{4} = 14\frac{1}{4}$ hours

4. Justin needs to take a taxi to the airport. It will cost \$1.75 for the first mile and \$0.50 for each additional $\frac{1}{4}$ of a mile. If he is 15 miles from the airport, how much will his taxi fare cost?

   $\frac{14}{1} \times \frac{4}{1} = \frac{56}{1} \times \frac{1}{2} = \$28$

   $\$28 + \$1.75 = \$29.75$

5. A rectangle has a length of $1\frac{1}{2}$ centimeters and a width of $\frac{3}{4}$ centimeters. Find its area and perimeter. A = l x w, P = 2(l) + 2(w)

   $A = \frac{3}{2} \times \frac{3}{4} = 1\frac{1}{8}$ centimeters

   $P = \frac{6}{2} + \frac{6}{4} = 4\frac{1}{2}$ centimeters

6. Selma needs $1\frac{1}{4}$ cups of corn syrup for a pecan pie. The bottle holds $2\frac{1}{2}$ cups. How much corn syrup will be left over?

   $\frac{10}{4} - \frac{5}{4} = 1\frac{1}{4}$ cups

Total Problems: Total Correct: Score:

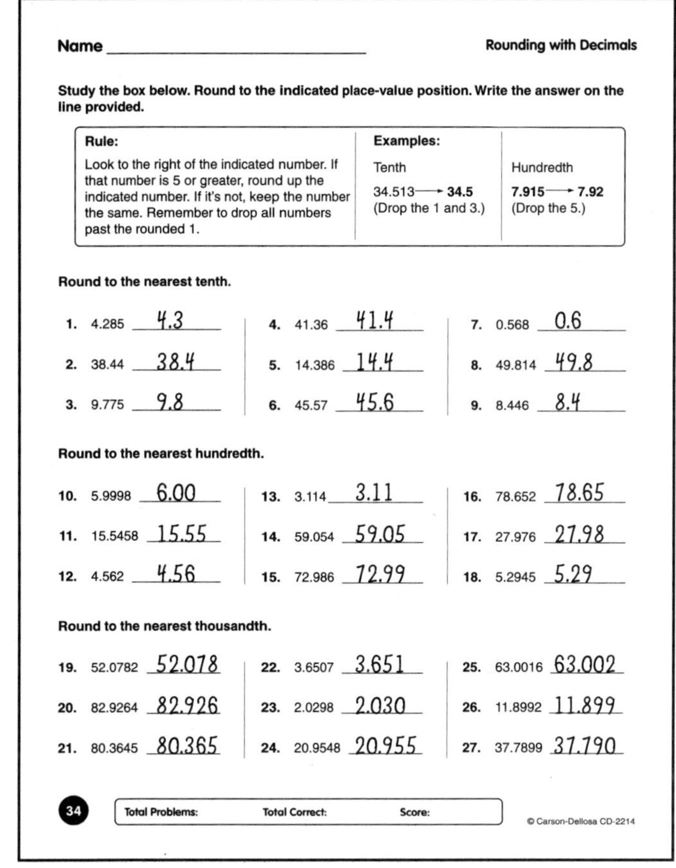

Name ____________________ **Rounding with Decimals**

Study the box below. Round to the indicated place-value position. Write the answer on the line provided.

Rule:	Examples:	
Look to the right of the indicated number. If that number is 5 or greater, round up the indicated number. If it's not, keep the number the same. Remember to drop all numbers past the rounded 1.	Tenth 34.513 → **34.5** (Drop the 1 and 3.)	Hundredth **7.915** → **7.92** (Drop the 5.)

Round to the nearest tenth.

1. 4.285 4.3
2. 38.44 38.4
3. 9.775 9.8
4. 41.36 41.4
5. 14.386 14.4
6. 45.57 45.6
7. 0.568 0.6
8. 49.814 49.8
9. 8.446 8.4

Round to the nearest hundredth.

10. 5.9998 6.00
11. 15.5458 15.55
12. 4.562 4.56
13. 3.114 3.11
14. 59.054 59.05
15. 72.986 72.99
16. 78.652 78.65
17. 27.976 27.98
18. 5.2945 5.29

Round to the nearest thousandth.

19. 52.0782 52.078
20. 82.9264 82.926
21. 80.3645 80.365
22. 3.6507 3.651
23. 2.0298 2.030
24. 20.9548 20.955
25. 63.0016 63.002
26. 11.8992 11.899
27. 37.7899 37.790

Total Problems: Total Correct: Score:

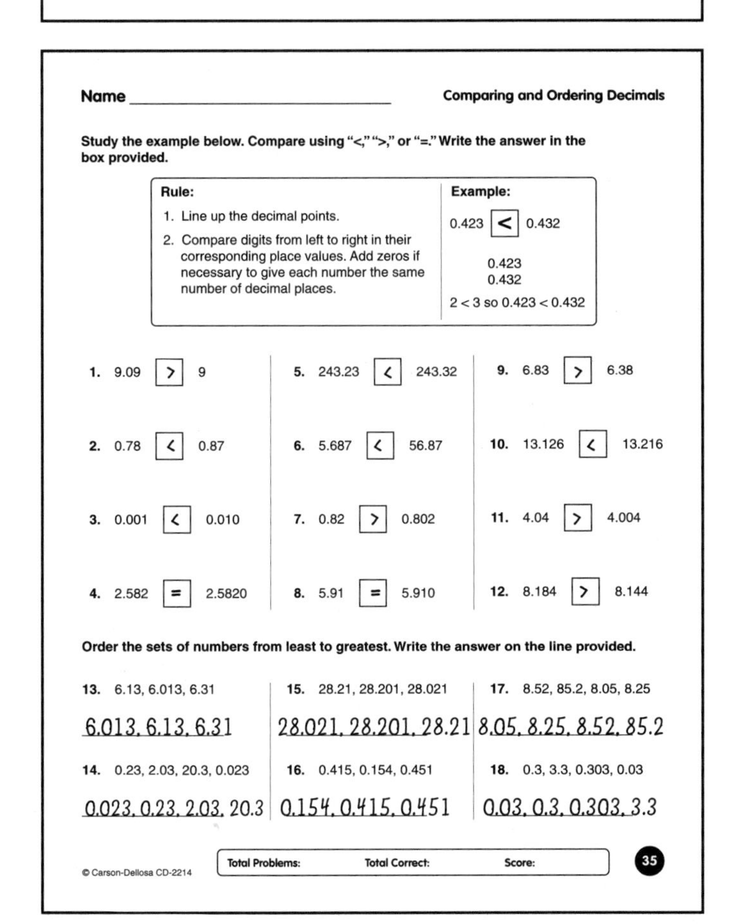

Name ____________________ **Comparing and Ordering Decimals**

Study the example below. Compare using "<," ">," or "=." Write the answer in the box provided.

Rule:	Example:
1. Line up the decimal points. 2. Compare digits from left to right in their corresponding place values. Add zeros if necessary to give each number the same number of decimal places.	0.423 [<] 0.432 0.423 0.432 2 < 3 so 0.423 < 0.432

1. 9.09 [>] 9
2. 0.78 [<] 0.87
3. 0.001 [<] 0.010
4. 2.582 [=] 2.5820
5. 243.23 [<] 243.32
6. 5.687 [<] 56.87
7. 0.82 [>] 0.802
8. 5.91 [=] 5.910
9. 6.83 [>] 6.38
10. 13.126 [<] 13.216
11. 4.04 [>] 4.004
12. 8.184 [>] 8.144

Order the sets of numbers from least to greatest. Write the answer on the line provided.

13. 6.13, 6.013, 6.31
    6.013, 6.13, 6.31
14. 0.23, 2.03, 20.3, 0.023
    0.023, 0.23, 2.03, 20.3
15. 28.21, 28.201, 28.021
    28.021, 28.201, 28.21
16. 0.415, 0.154, 0.451
    0.154, 0.415, 0.451
17. 8.52, 85.2, 8.05, 8.25
    8.05, 8.25, 8.52, 85.2
18. 0.3, 3.3, 0.303, 0.03
    0.03, 0.3, 0.303, 3.3

Total Problems: Total Correct: Score:

Name ____________________ **Adding Decimals**

Study the box below. Find each sum. Write the answer in the space provided.

Rule:	Example:
1. Line up the decimal points. 2. Add zeros if necessary. 3. Add as with whole numbers. 4. Carry (trade) if necessary. 5. Bring down the decimal point into the answer.	32.5 + 39.03 32.50 + 39.03 **71.53**

1. 0.71 + 1.62 = 2.33
2. 23.4267 + 34.5 = 57.9267
3. 431.51 + 273.68 = 705.19
4. 6.596 + 0.3 = 6.596 + 0.3 = 6.896
5. 23.73 + 62.83 = 86.56
6. 61.5 + 29.02 = 90.52
7. 645.9 + 481.765 = 1,127.665
8. 7.813 + 0.97 = 7.813 + 0.97 = 8.783
9. 363.4 + 340.8 = 704.2
10. 6.001 + 3.572 = 9.573
11. 24.093 + 7.819 = 31.912
12. 14 + 0.09 = 14 + 0.09 = 14.09
13. 815.15 + 105.07 = 920.22
14. 43.89 + 10.68 = 54.57
15. 12.708 + 43.91 = 56.618
16. 0.0059 + 3.2 = 0.0059 + 3.2 = 3.2059

Total Problems: Total Correct: Score:

Name ______________________________ **Subtracting Decimals**

Study the box below. Find each difference. Write the answer in the space provided.

Rule:	Example:
1. Line up the decimal points. 2. Add zeros if necessary. 3. Subtract as with whole numbers. 4. Borrow (trade) if necessary. 5. Bring down the decimal point into the answer.	9.05 – 6.347 9.050 – 6.347 **2.703**

1. 3.08 – 0.81 = 2.27
2. 28.026 – 6.7 = 21.326
3. 0.01 – 0.001 = 0.009
4. 0.57 – 0.08 = 0.49
5. 45.74 – 23.88 = 21.86
6. 62.3 – 17.06 = 45.24
7. 44.74 – 39.85 = 4.89
8. 16.52 – 0.49 = 16.03
9. 376.3 – 242.6 = 133.7
10. 4.005 – 2.606 = 1.399
11. 0.34 – 0.108 = 0.232
12. 16.5 – 15 = 1.5
13. 813.1 – 66.56 = 746.54
14. 64.8 – 9.89 = 54.91
15. 5.05 – 3.768 = 1.282
16. 20 – 8.745 = 11.255

Total Problems: Total Correct: Score:

Name ______________________________ **Estimating Sums and Differences with Decimals**

Study the box below. Estimate each sum or difference using one of the strategies presented in the box. Write the strategy used and the answer in the space provided.

Examples: Rounding	Clustering
23.62 → 24 – 6.71 → – 7 **17**	4.1 + 3.8 + 3.9 = 4 + 4 + 4 = 4 x 3 = **12**

1. 63.45 – 9.67 → 63 – 10 = 53 rounding
2. 18.72 + 15.95 → 19 + 16 = 35 rounding
3. 30.09 – 8.75 → 30 – 9 = 21 rounding
4. 3.4 + 2.8 + 3.3 → 9 rounding; 9 clustering
5. 27.38 – 6.51 → 27 – 7 = 20 rounding
6. 46.32 + 59.87 → 46 + 60 = 106 rounding
7. 56.25 + 78.91 → 56 + 79 = 135 rounding
8. 12.65 + 12.49 + 12.58 → 38 rounding; 36 clustering
9. 78.253 – 8.415 → 78 – 8 = 70 rounding
10. 40.63 + 59.79 → 41 + 60 = 101 rounding
11. 185.34 + 244.98 → 185 + 245 = 430 rounding
12. 29.82 + 31.04 + 28.88 → 90 rounding; 90 clustering
13. 270.31 – 148.85 → 270 – 149 = 121 rounding
14. 261.32 + 158.67 → 261 + 159 = 420 rounding
15. 67.15 – 27.77 → 67 – 28 = 39 rounding
16. 91.9 + 89.2 + 90.3 + 88.9 → 360 rounding; 360 clustering

Total Problems: Total Correct: Score:

Name ______________________________ **Multiplying Decimals**

Study the box below. Find each product and write the answer in the space provided.

Rule:	Example:
Multiply as you would whole numbers. The number of decimal places in the product is the sum of the decimal places in the factors. When the problem is presented horizontally, line up the numbers on the right. Do not line up the decimal points.	Factor → .23 — 2 decimal places Factor → x .5 — 1 decimal place Product → .115 — 3 decimal places

1. 1.4 x 9 = 12.6
2. 2.6 x 1.2 = 3.12
3. 20.3 x 0.06 = 1.218
4. 48.9 x 0.95 = 46.455
5. 4.52 x 0.9 = 4.068
6. 3.57 x 2.4 = 8.568
7. 1.002 x 3.5 = 3.507
8. 17.8 x 6.518 = 116.0204
9. 5.96 x 3.8 = 22.648
10. 6.3 x .29 = 1.827
11. 5.05 x 0.02 = 0.101
12. 9.6 x 35.56 = 341.376
13. 8.5 x .06 = 0.51
14. 16.4 x .75 = 12.3
15. 0.109 x 53.9 = 5.8751
16. 5.06 x 0.892 = 4.51352

Total Problems: Total Correct: Score:

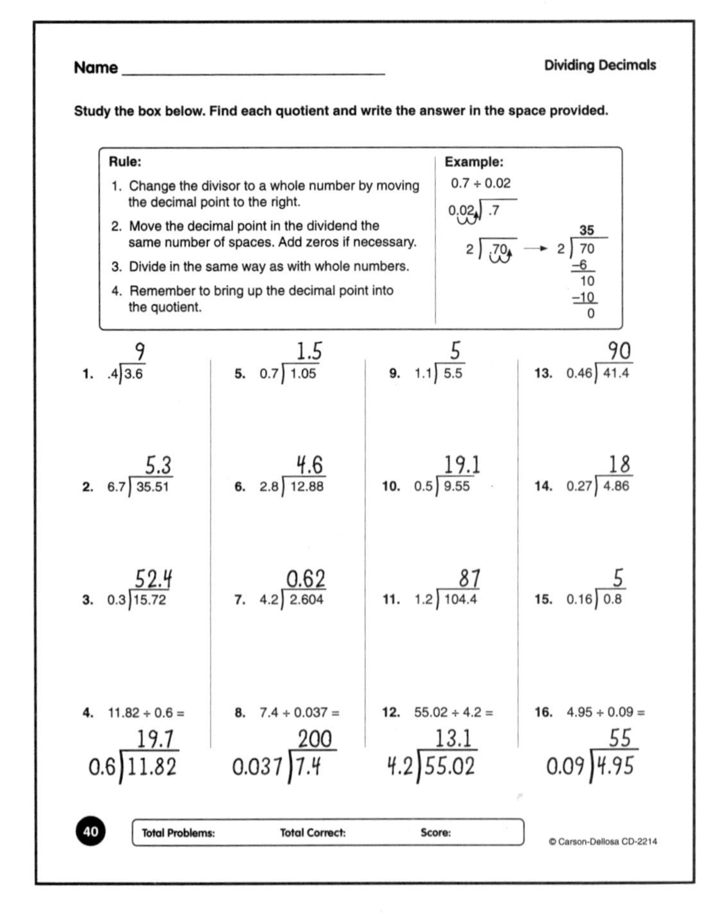

Name ______________________________ **Dividing Decimals**

Study the box below. Find each quotient and write the answer in the space provided.

Rule:	Example:
1. Change the divisor to a whole number by moving the decimal point to the right. 2. Move the decimal point in the dividend the same number of spaces. Add zeros if necessary. 3. Divide in the same way as with whole numbers. 4. Remember to bring up the decimal point into the quotient.	0.7 ÷ 0.02 0.02⟌.7 2⟌70 → 2⟌70 = 35 –6 10 –10 0

1. .4⟌3.6 = 9
2. 6.7⟌35.51 = 5.3
3. 0.3⟌15.72 = 52.4
4. 11.82 ÷ 0.6 = 19.7
5. 0.7⟌1.05 = 1.5
6. 2.8⟌12.88 = 4.6
7. 4.2⟌2.604 = 0.62
8. 7.4 ÷ 0.037 = 200
9. 1.1⟌5.5 = 5
10. 0.5⟌9.55 = 19.1
11. 1.2⟌104.4 = 87
12. 55.02 ÷ 4.2 = 13.1
13. 0.46⟌41.4 = 90
14. 0.27⟌4.86 = 18
15. 0.16⟌0.8 = 5
16. 4.95 ÷ 0.09 = 55

Total Problems: Total Correct: Score:

Name ______________________ **Estimating Products and Quotients with Decimals**

Study the examples below. Estimate each product or quotient using the strategy indicated below. When rounding, round to the highest place value. Write the answer on the line provided.

Examples:

Estimate by rounding.	Estimate using compatible numbers.
61.78 x 3.21	159.63 ÷ 3.91
60 x 3 = **180**	160 ÷ 4 = **40**

Estimate by rounding.

1. 6.92 x 8.45 = 7 x 8 = 56
2. 47.4 x 22.1 = 50 x 20 = 1,000
3. 29.36 x 4.08 = 30 x 4 = 120
4. 51.04 x 2.98 = 50 x 3 = 150
5. 5.832 x 9.5 = 6 x 10 = 60
6. 68.9 x 10.07 = 70 x 10 = 700
7. 0.002 x 0.03 = 0 x 0 = 0
8. 99.9 x 9.9 = 100 x 10 = 1,000
9. 11.246 x 6.8 = 10 x 7 = 70
10. 6.75 x 19.41 = 7 x 20 = 140
11. 42.9 x 21.87 = 40 x 20 = 800
12. 120.28 x 1.98 = 120 x 2 = 240

Estimate using compatible numbers.

13. 321.9 ÷ 8.2 = 320 ÷ 8 = 40
14. 299.3 ÷ 4.95 = 300 ÷ 5 = 60
15. 181.08 ÷ 30.9 = 180 ÷ 30 = 6
16. 1,241.5 ÷ 42.3 = 1,200 ÷ 40 = 30
17. 1,505.3 ÷ 30.4 = 1,500 ÷ 30 = 50
18. 243.6 ÷ 4.29 = 240 ÷ 4 = 60
19. 543.01 ÷ 9.2 = 540 ÷ 9 = 60
20. 636.7 ÷ 81 = 640 ÷ 80 = 8
21. 3,691.91 ÷ 6.2 = 3,600 ÷ 6 = 600
22. 4,495.6 ÷ 51 = 4,500 ÷ 50 = 90
23. 468.7 ÷ 6.2 = 480 ÷ 6 = 80
24. 492.51 ÷ 71.1 = 490 ÷ 70 = 7

Total Problems: Total Correct: Score:

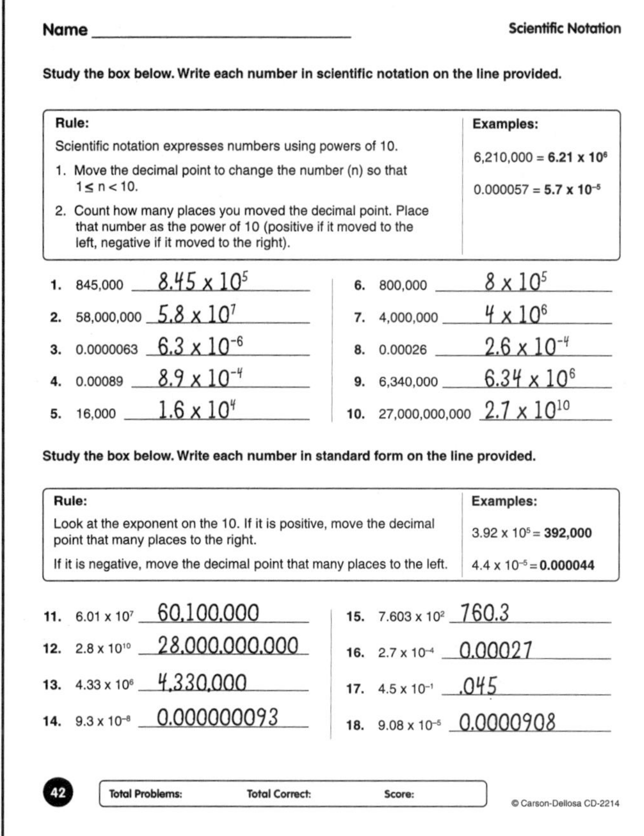

Name ______________________ **Scientific Notation**

Study the box below. Write each number in scientific notation on the line provided.

Rule:	Examples:
Scientific notation expresses numbers using powers of 10. 1. Move the decimal point to change the number (n) so that $1 \leq n < 10$. 2. Count how many places you moved the decimal point. Place that number as the power of 10 (positive if it moved to the left, negative if it moved to the right).	6,210,000 = **6.21×10^6** 0.000057 = **5.7×10^{-5}**

1. 845,000 8.45×10^5
2. 58,000,000 5.8×10^7
3. 0.0000063 6.3×10^{-6}
4. 0.00089 8.9×10^{-4}
5. 16,000 1.6×10^4
6. 800,000 8×10^5
7. 4,000,000 4×10^6
8. 0.00026 2.6×10^{-4}
9. 6,340,000 6.34×10^6
10. 27,000,000,000 2.7×10^{10}

Study the box below. Write each number in standard form on the line provided.

Rule:	Examples:
Look at the exponent on the 10. If it is positive, move the decimal point that many places to the right. If it is negative, move the decimal point that many places to the left.	3.92×10^5 = **392,000** 4.4×10^{-5} = **0.000044**

11. 6.01×10^7 60,100,000
12. 2.8×10^{10} 28,000,000,000
13. 4.33×10^6 4,330,000
14. 9.3×10^{-8} 0.000000093
15. 7.603×10^2 760.3
16. 2.7×10^{-4} 0.00027
17. 4.5×10^{-1} .045
18. 9.08×10^{-5} 0.0000908

Total Problems: Total Correct: Score:

Name ______________________ **Problem Solving with Decimals**

Solve each word problem. Show your work and write the answer in the space provided.

1. Janice, a piano teacher, charges $21.95 per lesson. If Sena has $200 to spend on lessons, how many will she be able to purchase?

   $21.95)$200 → 9 lessons

2. Mr. Wilson had to pay $5.75 for each member of his family to ride the train around the mountain. His total charge was $34.50. How many people did Mr. Wilson pay for?

   $5.75)$34.50 → 6 people

3. A magazine costs $2.79 at a newsstand. Terry orders a magazine subscription and pays $1.19 per issue. How much money is Terry saving per issue?

   $2.79 − $1.19 = $1.60

4. Tia ordered a hamburger for $1.99, large fries for $1.29, and a medium drink for $.89. How much did she spend?

   $1.99 + $1.29 + $ .89 = $4.17

5. Ivan earned $15.75 each week for mowing the lawn. If he spends $5.00 each week on ice cream and saves the rest, about how much will he have saved over a 12-week period?

   $15.75 − $ 5.00 = $10.75; $10.75 x 12 = $129.00

6. Brittany's class had a car wash to raise money for a trip. The students charged $5.25 per car and $7.50 per van. If they washed 26 cars and 18 vans, how much money did they make?

   $5.25 x 26 = $136.50; $7.50 x 18 = $135.00; $136.50 + $135.00 = $271.50

7. Mimi is saving money to buy a car. She has $3,459.81 in her account. If she can raise $5,000, her parents will match that amount. How much more money does she need to raise?

   $5,000.00 − $3,459.81 = $1,540.19

8. The following numbers are in order from least to greatest: 275, 307, 574, 889. Without changing the order of the numbers or adding any zeros, place decimal points in the numbers so that they are in order from greatest to least.

   275., 30.7, 5.74, .889

Total Problems: Total Correct: Score:

Name ______________________ **Ratios and Rates**

Study the box below. Express each ratio as a fraction in simplest form. Write the answer on the line provided.

Rule:	Example:
A ratio is a comparison of two numbers. It is often written as a fraction in simplest form.	8 to 10 $\frac{8}{10} = \frac{4}{5}$

1. 21 to 45 $\frac{7}{15}$
2. 6 to 72 $\frac{1}{12}$
3. 2 : 24 $\frac{1}{12}$
4. 30 to 15 $\frac{2}{1}$
5. 150 to 15 $\frac{10}{1}$
6. 5 out of 65 $\frac{1}{13}$
7. 64 : 18 $\frac{32}{9}$
8. 6 out of 39 $\frac{2}{13}$

Study the box below. Express each rate as a unit rate. Write the answer on the line provided.

Rules:	Example:
Rate: a ratio of 2 measurements with different units Unit rate: a rate in which the denominator is 1	$0.80 per 20 forks $0.80 ÷ 20 = **$0.04 per 1 fork**

9. $9.00 for 4 tapes — $2.25 for 1 tape
10. $150 for 2 days — $75 for 1 day
11. 330 miles in 5 hours — 66 miles in 1 hour
12. 3,000 miles in 4 days — 750 miles in 1 day
13. 16 people in 4 cars — 4 people in 1 car
14. $30 for 2 shirts — $15 for 1 shirt
15. 4,500 tickets in 2 days — 2,250 tickets in 1 day
16. 12 pounds in 6 weeks — 2 pounds in 1 week

Total Problems: Total Correct: Score:

Name ______________________ **Solving Proportions**

Study the box below. Solve each proportion. Write the answer on the line provided.

Rule:	Example:
A proportion is 2 equal ratios. To solve, find the cross products. Then, divide to solve for the variable.	$\frac{9}{15} = \frac{x}{10}$ $15x = 90$ $x = 90 \div 15$ **x = 6**

1. $\frac{3}{n} = \frac{5}{15}$ n = 9
2. $\frac{3}{5} = \frac{n}{25}$ n = 15
3. $\frac{3}{20} = \frac{n}{50}$ n = 7.5
4. $\frac{45}{20} = \frac{y}{4}$ y = 9
5. $\frac{6}{x} = \frac{42}{63}$ x = 9
6. $\frac{8}{5} = \frac{6}{x}$ x = 3.75
7. $\frac{4}{5} = \frac{d}{6}$ d = 4.8
8. $\frac{17}{8.5} = \frac{n}{0.01}$ n = .02
9. $\frac{n}{8} = \frac{12}{16}$ n = 6
10. $\frac{13}{4} = \frac{52}{y}$ y = 16
11. $\frac{12}{w} = \frac{8}{14}$ w = 21
12. $\frac{60}{75} = \frac{r}{5}$ r = 4
13. $\frac{x}{2.4} = \frac{4}{0.6}$ x = 16
14. $\frac{18}{m} = \frac{300}{24}$ m = 1.44
15. $\frac{2}{m} = \frac{5}{8.75}$ m = 3.5
16. $\frac{0.24}{y} = \frac{3}{9.6}$ y = .768
17. $\frac{8}{35} = \frac{n}{350}$ n = 80
18. $\frac{x}{4} = \frac{18}{30}$ x = 2.4
19. $\frac{x}{40} = \frac{15}{24}$ x = 25
20. $\frac{5}{1.6} = \frac{m}{4.8}$ m = 15
21. $\frac{1}{2} = \frac{x}{4.2}$ x = 2.1
22. $\frac{p}{5} = \frac{450}{4.5}$ p = 500
23. $\frac{7}{4} = \frac{n}{0.8}$ n = 1.4
24. $\frac{0.1}{8.2} = \frac{1.8}{x}$ x = 147.6

Total Problems: Total Correct: Score:

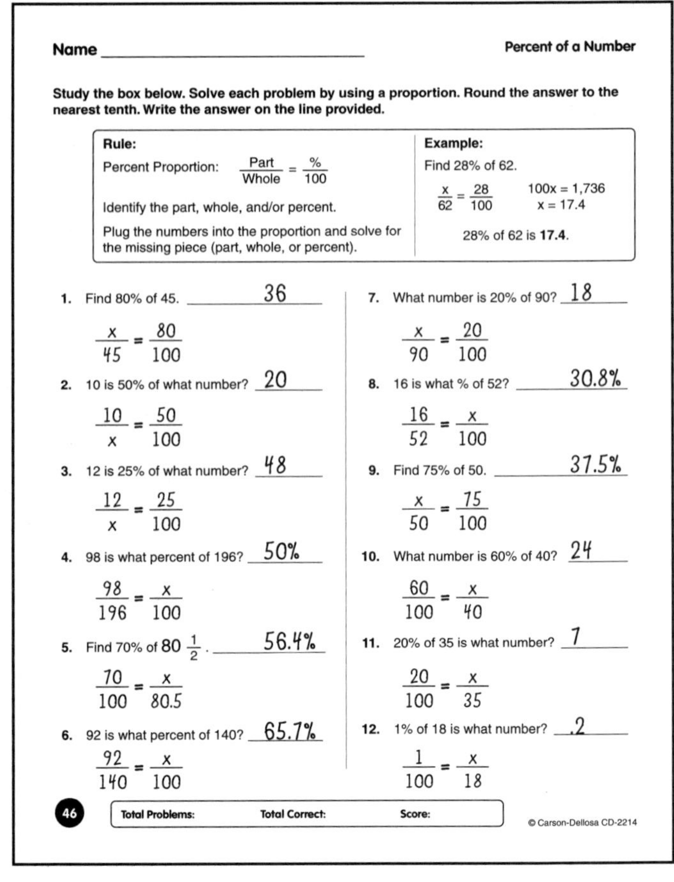

Name ______________________ **Percent of a Number**

Study the box below. Solve each problem by using a proportion. Round the answer to the nearest tenth. Write the answer on the line provided.

Rule:	Example:
Percent Proportion: $\frac{Part}{Whole} = \frac{\%}{100}$ Identify the part, whole, and/or percent. Plug the numbers into the proportion and solve for the missing piece (part, whole, or percent).	Find 28% of 62. $\frac{x}{62} = \frac{28}{100}$ $100x = 1{,}736$, $x = 17.4$ 28% of 62 is **17.4**.

1. Find 80% of 45. 36 — $\frac{x}{45} = \frac{80}{100}$
2. 10 is 50% of what number? 20 — $\frac{10}{x} = \frac{50}{100}$
3. 12 is 25% of what number? 48 — $\frac{12}{x} = \frac{25}{100}$
4. 98 is what percent of 196? 50% — $\frac{98}{196} = \frac{x}{100}$
5. Find 70% of $80\frac{1}{2}$. 56.4% — $\frac{70}{100} = \frac{x}{80.5}$
6. 92 is what percent of 140? 65.7% — $\frac{92}{140} = \frac{x}{100}$
7. What number is 20% of 90? 18 — $\frac{x}{90} = \frac{20}{100}$
8. 16 is what % of 52? 30.8% — $\frac{16}{52} = \frac{x}{100}$
9. Find 75% of 50. 37.5% — $\frac{x}{50} = \frac{75}{100}$
10. What number is 60% of 40? 24 — $\frac{60}{100} = \frac{x}{40}$
11. 20% of 35 is what number? 7 — $\frac{20}{100} = \frac{x}{35}$
12. 1% of 18 is what number? .2 — $\frac{1}{100} = \frac{x}{18}$

Total Problems: Total Correct: Score:

Name ______________________ **Percent of Change**

Study the box below. Find the percent of increase or decrease. Round to the nearest whole percent. Write the answer on the line provided.

Rule:	Example:
1. Subtract to find the amount of increase or decrease. 2. Write a proportion: $\frac{change}{original} = \frac{x\%}{100}$ 3. Cross multiply to find the percent of change.	old: 29 (original) new: 32 change: 32 – 29 = 3 $\frac{3}{29} = \frac{x}{100}$ $300 = 29x$ **x = 10%**

1. old: \$11, new: \$9 — 18% — $\frac{2}{11} = \frac{x}{100}$
2. old: 14, new: 20 — 43% — $\frac{6}{14} = \frac{x}{100}$
3. old: 50, new: 40 — 20% — $\frac{10}{50} = \frac{x}{100}$
4. old: 20, new: 30 — 50% — $\frac{10}{20} = \frac{x}{100}$
5. old: \$80, new: \$40 — 50% — $\frac{40}{80} = \frac{x}{100}$
6. old: \$4.50, new: \$5.90 — 31% — $\frac{1.40}{4.50} = \frac{x}{100}$
7. old: \$278, new: \$350 — 26% — $\frac{72}{278} = \frac{x}{100}$
8. old: 70, new: 35 — 50% — $\frac{35}{70} = \frac{x}{100}$
9. old: 500, new: 300 — 40% — $\frac{200}{500} = \frac{x}{100}$
10. old: \$179, new: \$211 — 18% — $\frac{32}{179} = \frac{x}{100}$
11. old: 175, new: 149 — 15% — $\frac{26}{175} = \frac{x}{100}$
12. old: \$550, new: \$450 — 18% — $\frac{100}{550} = \frac{x}{100}$
13. old: 91, new: 89 — 2% — $\frac{2}{91} = \frac{x}{100}$
14. old: 0.67, new: 1.92 — 187% — $\frac{1.25}{0.67} = \frac{x}{100}$
15. old: 8.2, new: 10 — 22% — $\frac{1.8}{8.2} = \frac{x}{100}$
16. old: 72, new: 60 — 17% — $\frac{12}{72} = \frac{x}{100}$
17. old: \$126, new: \$150 — 19% — $\frac{24}{126} = \frac{x}{100}$
18. old: 0.95, new: 1.5 — 58% — $\frac{.55}{.95} = \frac{x}{100}$

Total Problems: Total Correct: Score:

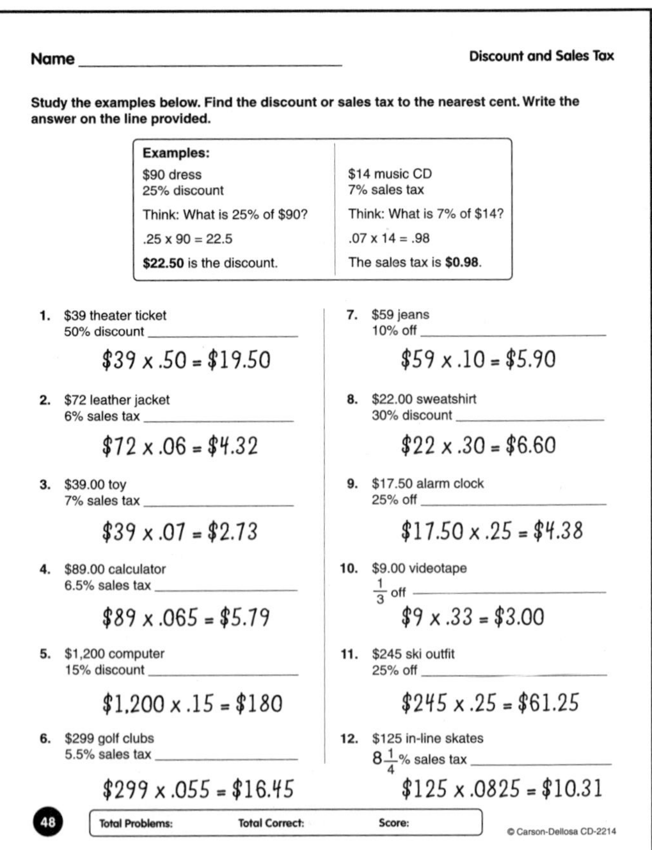

Name ______________________ **Discount and Sales Tax**

Study the examples below. Find the discount or sales tax to the nearest cent. Write the answer on the line provided.

Examples:	
\$90 dress 25% discount Think: What is 25% of \$90? .25 x 90 = 22.5 **\$22.50** is the discount.	\$14 music CD 7% sales tax Think: What is 7% of \$14? .07 x 14 = .98 The sales tax is **\$0.98**.

1. \$39 theater ticket, 50% discount — \$39 x .50 = \$19.50
2. \$72 leather jacket, 6% sales tax — \$72 x .06 = \$4.32
3. \$39.00 toy, 7% sales tax — \$39 x .07 = \$2.73
4. \$89.00 calculator, 6.5% sales tax — \$89 x .065 = \$5.79
5. \$1,200 computer, 15% discount — \$1,200 x .15 = \$180
6. \$299 golf clubs, 5.5% sales tax — \$299 x .055 = \$16.45
7. \$59 jeans, 10% off — \$59 x .10 = \$5.90
8. \$22.00 sweatshirt, 30% discount — \$22 x .30 = \$6.60
9. \$17.50 alarm clock, 25% off — \$17.50 x .25 = \$4.38
10. \$9.00 videotape, $\frac{1}{3}$ off — \$9 x .33 = \$3.00
11. \$245 ski outfit, 25% off — \$245 x .25 = \$61.25
12. \$125 in-line skates, $8\frac{1}{4}$% sales tax — \$125 x .0825 = \$10.31

Total Problems: Total Correct: Score:

Name ______________________________ **Percents and Fractions**

Study the box below. Write each fraction as a percent. Round to the nearest hundredth. Write the answer on the line provided.

Rule:	Example:
To change the fraction to a percent:	$\frac{7}{8}$
1. Divide the numerator by the denominator.	
2. Change the decimal to a percent. Move the decimal point 2 places to the right. Add a 0 if necessary, and do not forget the percent sign.	$7 \div 8 = .875$ **87.5%**

1. $\frac{11}{100}$ 11%
2. $\frac{7}{14}$ 50%
3. $\frac{5}{8}$ 62.5%
4. $\frac{16}{25}$ 64%
5. $\frac{1}{4}$ 25%
6. $\frac{20}{120}$ 16.67%
7. $\frac{1}{10}$ 10%
8. $\frac{1}{3}$ 33.33%
9. $\frac{106}{100}$ 106%
10. $\frac{9}{20}$ 45%
11. $\frac{5}{2}$ 250%
12. $1\frac{1}{2}$ 150%

Study the box below. Write each percent as a fraction or mixed number in lowest terms. Write the answer on the line provided.

Rule:	Example:
To change a percent to a fraction:	$40\% = \frac{40}{100} = \frac{2}{5}$
1. Drop the percent sign and place the number over 100.	
2. Reduce the fraction to lowest terms.	

13. 30% = $\frac{3}{10}$
14. 1% = $\frac{1}{100}$
15. 75% = $\frac{3}{4}$
16. 31% = $\frac{31}{100}$
17. 67.5% = $\frac{27}{40}$
18. 99% = $\frac{99}{100}$
19. 0.3% = $\frac{3}{1,000}$
20. 19% = $\frac{19}{100}$

Total Problems:	Total Correct:	Score:

Name ______________________________ **Percents and Decimals**

Study the box below. Express each decimal as a percent. Write the answer on the line provided.

Rule:	Example:
To change a decimal to a percent: Multiply by 100 (which moves the decimal point 2 places to the right) and add the percent sign.	0.7 0.7 x 100 = **70%**

1. 0.42 = 42%
2. 0.4 = 40%
3. 0.52 = 52%
4. 0.25 = 25%
5. 0.02 = 2%
6. 0.05 = 5%
7. 0.5 = 50%
8. 0.92 = 92%
9. 0.565 = 56.5%
10. 0.29 = 29%
11. 0.875 = 87.5%
12. 1.42 = 142%

Study the box below. Express each percent as a decimal. Write the answer on the line provided.

Rule:	Examples:	
To change a percent to a decimal: Divide by 100 (which moves the decimal point 2 places to the left).	71% 71 ÷ 100 = **.71**	8% 8 ÷ 100 = **.08**

13. 32% = .32
14. 4.5% = .045
15. 200% = 2
16. 61% = .61
17. 81.2% = .812
18. 0.1% = .001
19. 8% = .08
20. 70% = .7
21. 100% = 1
22. 1% = .01
23. 12% = .12
24. 0.05% = .0005

Total Problems:	Total Correct:	Score:

Name ______________________________ **Problem Solving with Ratio, Proportion, and Percent**

Solve each word problem. Show your work and write the answer in the space provided.

1. Harrison earned $365 in commissions during the first week of the year by selling refrigerators. This was 18% of his total sales for the month. What were his total sales for the month?

   $\frac{18}{100} = \frac{365}{x}$ $2,027.78

2. Maria sold 28 dolls over the weekend at her toy store. This was 30% of all the dolls she had in stock. How many dolls were in stock?

   $\frac{30}{100} = \frac{28}{x}$ 93 dolls

3. Britt's team won 70% of their games. They played 20 games during the season. How many games did they win?

   $\frac{70}{100} = \frac{x}{20}$ 14 games

4. In the final game of Lana's season, 12 girls played. If 90% of the team played in that game, how many girls were on the team?

   $\frac{90}{100} = \frac{12}{x}$ 13 girls

5. The cheerleaders ordered 250 spirit ribbons to sell at each basketball game. At the first game, they sold 179. What percent of the ribbons did they sell?

   $\frac{179}{250} = \frac{x}{100}$ 72% of the ribbons

6. Refer to problem 5. During the second game, the cheerleaders sold 215 spirit ribbons. Find the percent of increase.

   $\frac{18}{100} = \frac{365}{x}$ 20% increase

7. Cory traveled 18 miles in his car on 1 gallon of gasoline. At that same rate of fuel consumption, how far would he be able to go on 12 gallons?

   18 x 12 = 216 miles

8. Mollie was traveling in Israel on vacation. She purchased a souvenir for 65 shekels (Israeli currency). If the exchange rate was 3 shekels : 1 dollar, how many dollars did the souvenir cost?

   $\frac{3}{1} = \frac{65}{x}$ $21.67

Total Problems:	Total Correct:	Score:

Name ______________________________ **Writing Algebraic Expressions and Equations**

Study the box below. On the line provided, write an algebraic expression or equation for each phrase.

Example:	Tip:
Three times the cost of the sweater is $60. Let c represent the cost of the sweater. The word "times" suggests multiplication, and "is" means an equal sign. **3c = 60**	Look for the key words which indicate addition, subtraction, multiplication, division, or equals.

1. the product of r and 9 — 9r
2. 12 more than y — y + 12
3. the quotient of w and 25 — w ÷ 25
4. x subtracted from 21 — 21 – x
5. m divided by 18 — m ÷ 18
6. One-third of n is 12. — n ÷ 3 = 12
7. Twice a number is 28. — 2n = 28
8. Two more than the quotient of 6 and x is 5. — 6 ÷ x + 2 = 5
9. Five increased by n is 10. — 5 + n = 10
10. the sum of b and 16 — b + 16
11. the difference of y and 5 — y – 5
12. 16 less than e — e – 16
13. the product of d and 15 — 15d
14. Eleven times y is 22. — 11y = 22
15. The sum of y and 5 is 32. — y + 5 = 32
16. Eight decreased by a number is 3. — 8 – n = 3
17. Fifteen decreased by the product of 9 and a number is 12. — 15 – 9n = 12
18. Three less than the sum of 5 and y is 10. — (5 + y) – 3 = 10

Total Problems:	Total Correct:	Score:

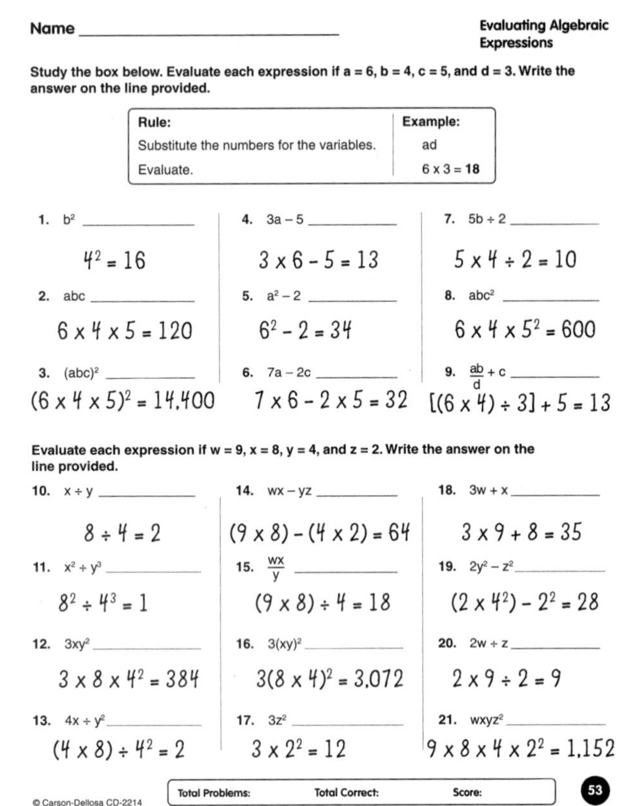

Name ____________________ **Evaluating Algebraic Expressions**

Study the box below. Evaluate each expression if a = 6, b = 4, c = 5, and d = 3. Write the answer on the line provided.

Rule:	Example:
Substitute the numbers for the variables.	ad
Evaluate.	6 x 3 = **18**

1. b^2 — $4^2 = 16$
2. abc — $6 \times 4 \times 5 = 120$
3. $(abc)^2$ — $(6 \times 4 \times 5)^2 = 14{,}400$
4. 3a – 5 — $3 \times 6 - 5 = 13$
5. $a^2 - 2$ — $6^2 - 2 = 34$
6. 7a – 2c — $7 \times 6 - 2 \times 5 = 32$
7. 5b ÷ 2 — $5 \times 4 \div 2 = 10$
8. abc^2 — $6 \times 4 \times 5^2 = 600$
9. $\frac{ab}{d} + c$ — $[(6 \times 4) \div 3] + 5 = 13$

Evaluate each expression if w = 9, x = 8, y = 4, and z = 2. Write the answer on the line provided.

10. x ÷ y — $8 \div 4 = 2$
11. $x^2 \div y^3$ — $8^2 \div 4^3 = 1$
12. $3xy^2$ — $3 \times 8 \times 4^2 = 384$
13. $4x \div y^2$ — $(4 \times 8) \div 4^2 = 2$
14. wx – yz — $(9 \times 8) - (4 \times 2) = 64$
15. $\frac{wx}{y}$ — $(9 \times 8) \div 4 = 18$
16. $3(xy)^2$ — $3(8 \times 4)^2 = 3{,}072$
17. $3z^2$ — $3 \times 2^2 = 12$
18. 3w + x — $3 \times 9 + 8 = 35$
19. $2y^2 - z^2$ — $(2 \times 4^2) - 2^2 = 28$
20. 2w ÷ z — $2 \times 9 \div 2 = 9$
21. $wxyz^2$ — $9 \times 8 \times 4 \times 2^2 = 1{,}152$

© Carson-Dellosa CD-2214 | Total Problems: | Total Correct: | Score: | 53

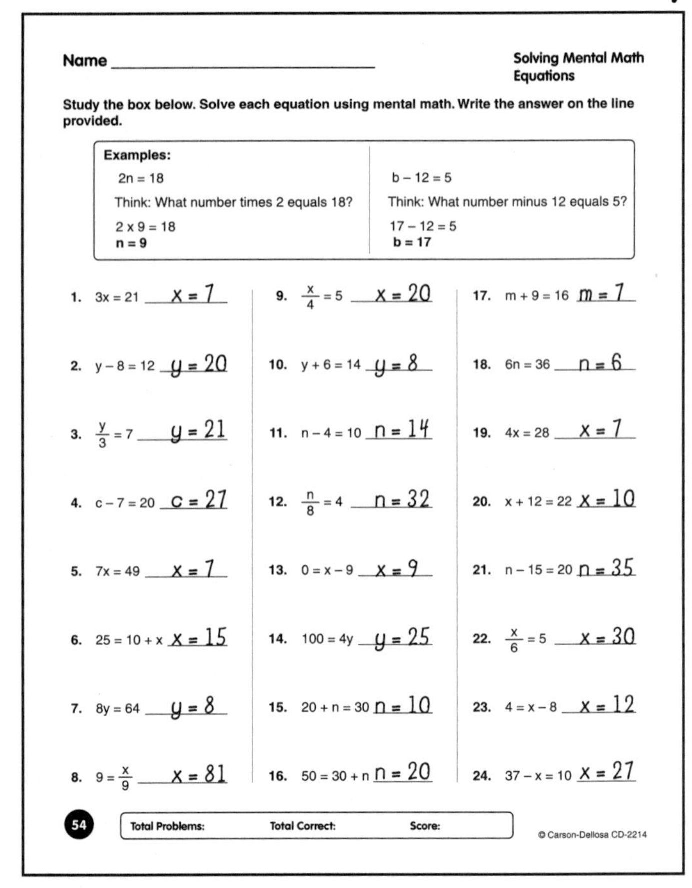

Name ____________________ **Solving Mental Math Equations**

Study the box below. Solve each equation using mental math. Write the answer on the line provided.

Examples:	
2n = 18	b – 12 = 5
Think: What number times 2 equals 18?	Think: What number minus 12 equals 5?
2 x 9 = 18 **n = 9**	17 – 12 = 5 **b = 17**

1. 3x = 21 — x = 7
2. y – 8 = 12 — y = 20
3. $\frac{y}{3} = 7$ — y = 21
4. c – 7 = 20 — c = 27
5. 7x = 49 — x = 7
6. 25 = 10 + x — x = 15
7. 8y = 64 — y = 8
8. $9 = \frac{x}{9}$ — x = 81
9. $\frac{x}{4} = 5$ — x = 20
10. y + 6 = 14 — y = 8
11. n – 4 = 10 — n = 14
12. $\frac{n}{8} = 4$ — n = 32
13. 0 = x – 9 — x = 9
14. 100 = 4y — y = 25
15. 20 + n = 30 — n = 10
16. 50 = 30 + n — n = 20
17. m + 9 = 16 — m = 7
18. 6n = 36 — n = 6
19. 4x = 28 — x = 7
20. x + 12 = 22 — x = 10
21. n – 15 = 20 — n = 35
22. $\frac{x}{6} = 5$ — x = 30
23. 4 = x – 8 — x = 12
24. 37 – x = 10 — x = 27

54 | Total Problems: | Total Correct: | Score: | © Carson-Dellosa CD-2214

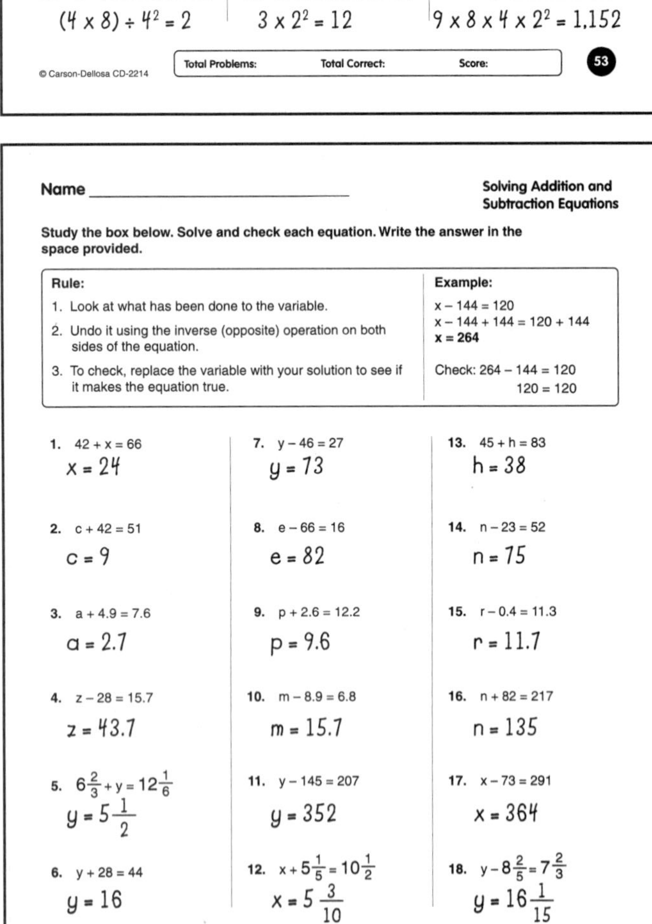

Name ____________________ **Solving Addition and Subtraction Equations**

Study the box below. Solve and check each equation. Write the answer in the space provided.

Rule:	Example:
1. Look at what has been done to the variable.	x – 144 = 120
2. Undo it using the inverse (opposite) operation on both sides of the equation.	x – 144 + 144 = 120 + 144 **x = 264**
3. To check, replace the variable with your solution to see if it makes the equation true.	Check: 264 – 144 = 120 120 = 120

1. 42 + x = 66 — x = 24
2. c + 42 = 51 — c = 9
3. a + 4.9 = 7.6 — a = 2.7
4. z – 28 = 15.7 — z = 43.7
5. $6\frac{2}{3} + y = 12\frac{1}{6}$ — $y = 5\frac{1}{2}$
6. y + 28 = 44 — y = 16
7. y – 46 = 27 — y = 73
8. e – 66 = 16 — e = 82
9. p + 2.6 = 12.2 — p = 9.6
10. m – 8.9 = 6.8 — m = 15.7
11. y – 145 = 207 — y = 352
12. $x + 5\frac{1}{5} = 10\frac{1}{2}$ — $x = 5\frac{3}{10}$
13. 45 + h = 83 — h = 38
14. n – 23 = 52 — n = 75
15. r – 0.4 = 11.3 — r = 11.7
16. n + 82 = 217 — n = 135
17. x – 73 = 291 — x = 364
18. $y - 8\frac{2}{5} = 7\frac{2}{3}$ — $y = 16\frac{1}{15}$

© Carson-Dellosa CD-2214 | Total Problems: | Total Correct: | Score: | 55

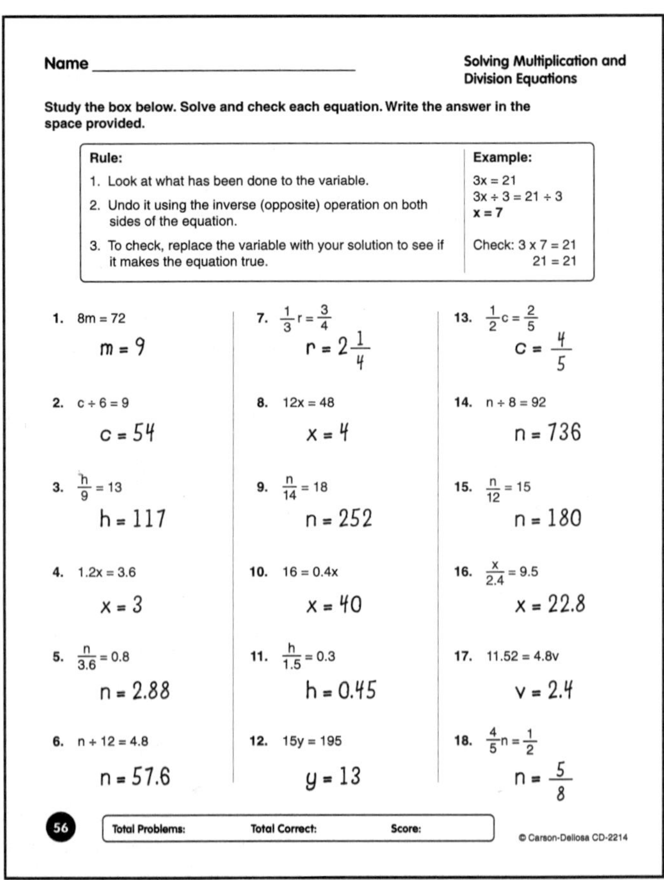

Name ____________________ **Solving Multiplication and Division Equations**

Study the box below. Solve and check each equation. Write the answer in the space provided.

Rule:	Example:
1. Look at what has been done to the variable.	3x = 21
2. Undo it using the inverse (opposite) operation on both sides of the equation.	3x ÷ 3 = 21 ÷ 3 **x = 7**
3. To check, replace the variable with your solution to see if it makes the equation true.	Check: 3 x 7 = 21 21 = 21

1. 8m = 72 — m = 9
2. c ÷ 6 = 9 — c = 54
3. $\frac{h}{9} = 13$ — h = 117
4. 1.2x = 3.6 — x = 3
5. $\frac{n}{3.6} = 0.8$ — n = 2.88
6. n ÷ 12 = 4.8 — n = 57.6
7. $\frac{1}{3}r = \frac{3}{4}$ — $r = 2\frac{1}{4}$
8. 12x = 48 — x = 4
9. $\frac{n}{14} = 18$ — n = 252
10. 16 = 0.4x — x = 40
11. $\frac{h}{1.5} = 0.3$ — h = 0.45
12. 15y = 195 — y = 13
13. $\frac{1}{2}c = \frac{2}{5}$ — $c = \frac{4}{5}$
14. n ÷ 8 = 92 — n = 736
15. $\frac{n}{12} = 15$ — n = 180
16. $\frac{x}{2.4} = 9.5$ — x = 22.8
17. 11.52 = 4.8v — v = 2.4
18. $\frac{4}{5}n = \frac{1}{2}$ — $n = \frac{5}{8}$

56 | Total Problems: | Total Correct: | Score: | © Carson-Dellosa CD-2214

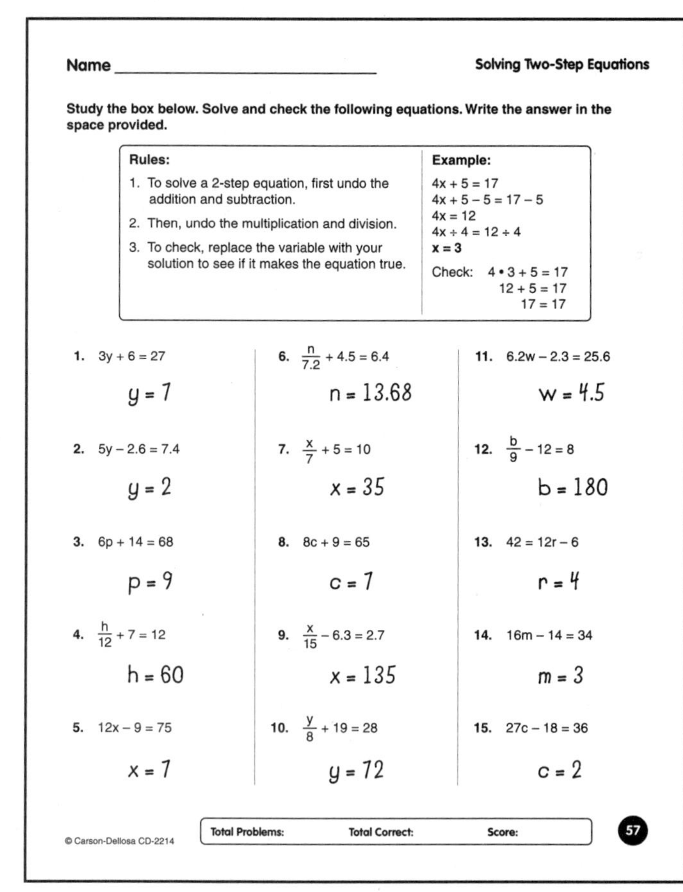

Name ______________________ **Solving Two-Step Equations**

Study the box below. Solve and check the following equations. Write the answer in the space provided.

Rules:	Example:
1. To solve a 2-step equation, first undo the addition and subtraction. 2. Then, undo the multiplication and division. 3. To check, replace the variable with your solution to see if it makes the equation true.	$4x + 5 = 17$ $4x + 5 - 5 = 17 - 5$ $4x = 12$ $4x \div 4 = 12 \div 4$ **$x = 3$** Check: $4 \cdot 3 + 5 = 17$ $12 + 5 = 17$ $17 = 17$

1. $3y + 6 = 27$ — $y = 7$
2. $5y - 2.6 = 7.4$ — $y = 2$
3. $6p + 14 = 68$ — $p = 9$
4. $\frac{h}{12} + 7 = 12$ — $h = 60$
5. $12x - 9 = 75$ — $x = 7$
6. $\frac{n}{7.2} + 4.5 = 6.4$ — $n = 13.68$
7. $\frac{x}{7} + 5 = 10$ — $x = 35$
8. $8c + 9 = 65$ — $c = 7$
9. $\frac{x}{15} - 6.3 = 2.7$ — $x = 135$
10. $\frac{y}{8} + 19 = 28$ — $y = 72$
11. $6.2w - 2.3 = 25.6$ — $w = 4.5$
12. $\frac{b}{9} - 12 = 8$ — $b = 180$
13. $42 = 12r - 6$ — $r = 4$
14. $16m - 14 = 34$ — $m = 3$
15. $27c - 18 = 36$ — $c = 2$

© Carson-Dellosa CD-2214 Total Problems: Total Correct: Score: 57

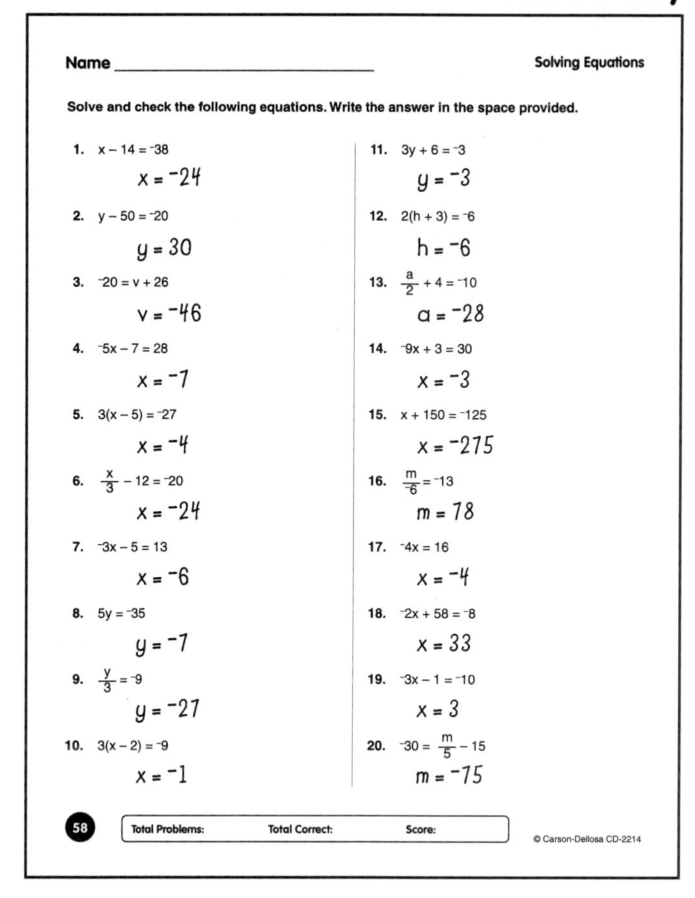

Name ______________________ **Solving Equations**

Solve and check the following equations. Write the answer in the space provided.

1. $x - 14 = {}^-38$ — $x = {}^-24$
2. $y - 50 = {}^-20$ — $y = 30$
3. ${}^-20 = v + 26$ — $v = {}^-46$
4. ${}^-5x - 7 = 28$ — $x = {}^-7$
5. $3(x - 5) = {}^-27$ — $x = {}^-4$
6. $\frac{x}{3} - 12 = {}^-20$ — $x = {}^-24$
7. ${}^-3x - 5 = 13$ — $x = {}^-6$
8. $5y = {}^-35$ — $y = {}^-7$
9. $\frac{y}{3} = {}^-9$ — $y = {}^-27$
10. $3(x - 2) = {}^-9$ — $x = {}^-1$
11. $3y + 6 = {}^-3$ — $y = {}^-3$
12. $2(h + 3) = {}^-6$ — $h = {}^-6$
13. $\frac{a}{2} + 4 = {}^-10$ — $a = {}^-28$
14. ${}^-9x + 3 = 30$ — $x = {}^-3$
15. $x + 150 = {}^-125$ — $x = {}^-275$
16. $\frac{m}{{}^-6} = {}^-13$ — $m = 78$
17. ${}^-4x = 16$ — $x = {}^-4$
18. ${}^-2x + 58 = {}^-8$ — $x = 33$
19. ${}^-3x - 1 = {}^-10$ — $x = 3$
20. ${}^-30 = \frac{m}{5} - 15$ — $m = {}^-75$

58 Total Problems: Total Correct: Score: © Carson-Dellosa CD-2214

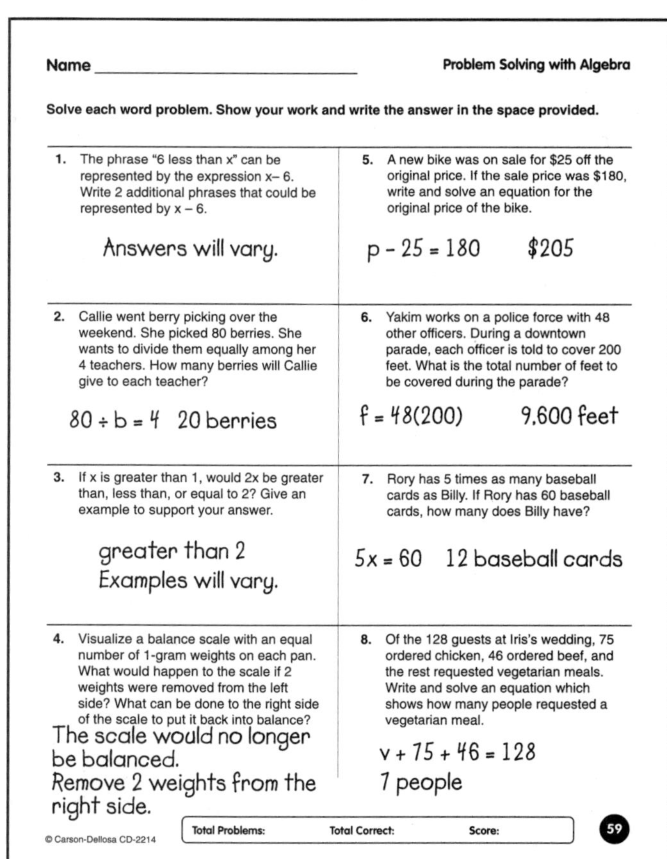

Name ______________________ **Problem Solving with Algebra**

Solve each word problem. Show your work and write the answer in the space provided.

1. The phrase "6 less than x" can be represented by the expression x– 6. Write 2 additional phrases that could be represented by x – 6.
   Answers will vary.
2. Callie went berry picking over the weekend. She picked 80 berries. She wants to divide them equally among her 4 teachers. How many berries will Callie give to each teacher?
   $80 \div b = 4$ 20 berries
3. If x is greater than 1, would 2x be greater than, less than, or equal to 2? Give an example to support your answer.
   greater than 2
   Examples will vary.
4. Visualize a balance scale with an equal number of 1-gram weights on each pan. What would happen to the scale if 2 weights were removed from the left side? What can be done to the right side of the scale to put it back into balance?
   The scale would no longer be balanced.
   Remove 2 weights from the right side.
5. A new bike was on sale for $25 off the original price. If the sale price was $180, write and solve an equation for the original price of the bike.
   $p - 25 = 180$ $205
6. Yakim works on a police force with 48 other officers. During a downtown parade, each officer is told to cover 200 feet. What is the total number of feet to be covered during the parade?
   $f = 48(200)$ 9,600 feet
7. Rory has 5 times as many baseball cards as Billy. If Rory has 60 baseball cards, how many does Billy have?
   $5x = 60$ 12 baseball cards
8. Of the 128 guests at Iris's wedding, 75 ordered chicken, 46 ordered beef, and the rest requested vegetarian meals. Write and solve an equation which shows how many people requested a vegetarian meal.
   $v + 75 + 46 = 128$
   7 people

© Carson-Dellosa CD-2214 Total Problems: Total Correct: Score: 59

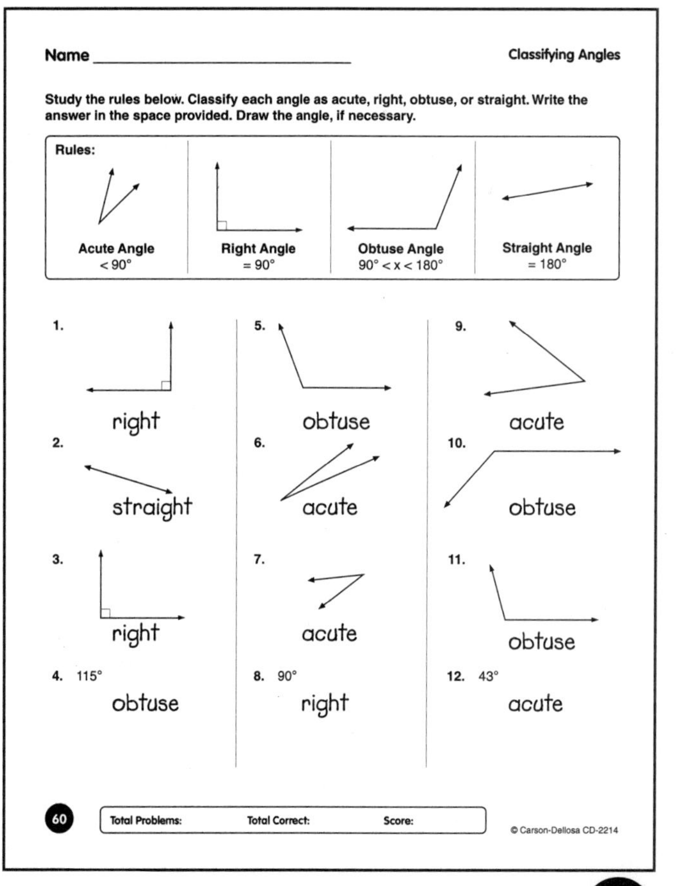

Name ______________________ **Classifying Angles**

Study the rules below. Classify each angle as acute, right, obtuse, or straight. Write the answer in the space provided. Draw the angle, if necessary.

Rules:

Acute Angle	Right Angle	Obtuse Angle	Straight Angle
< 90°	= 90°	90° < x < 180°	= 180°

1. right
2. straight
3. right
4. 115° obtuse
5. obtuse
6. acute
7. acute
8. 90° right
9. acute
10. obtuse
11. obtuse
12. 43° acute

60 Total Problems: Total Correct: Score: © Carson-Dellosa CD-2214

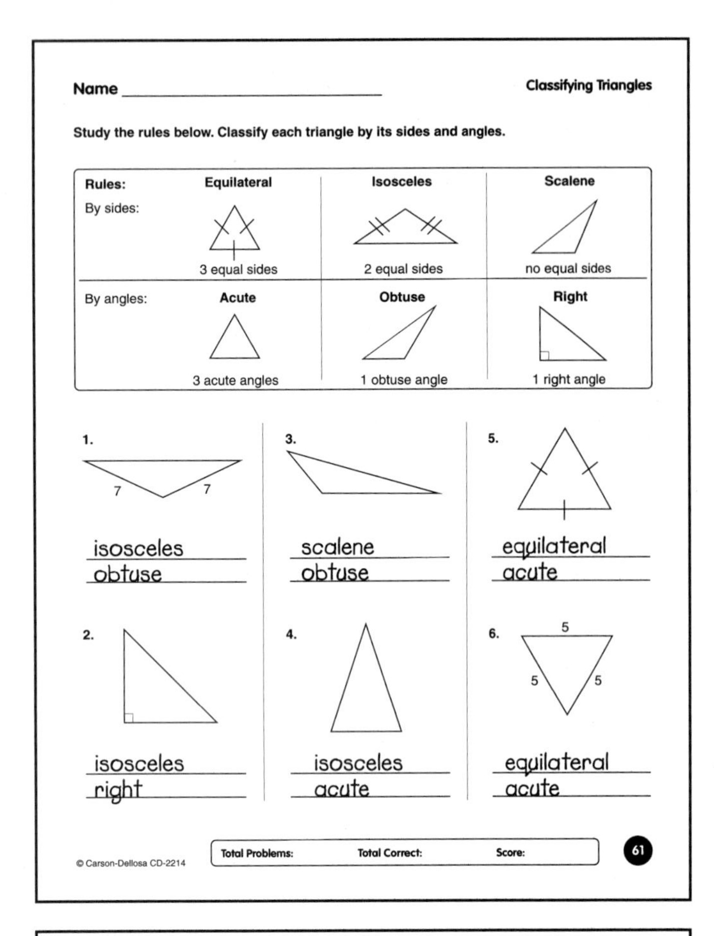

Name ____________________ **Classifying Triangles**

Study the rules below. Classify each triangle by its sides and angles.

Rules:	Equilateral	Isosceles	Scalene
By sides:	3 equal sides	2 equal sides	no equal sides
By angles:	**Acute** 3 acute angles	**Obtuse** 1 obtuse angle	**Right** 1 right angle

1. (7, 7) isosceles, obtuse
2. isosceles, right
3. scalene, obtuse
4. isosceles, acute
5. equilateral, acute
6. (5, 5, 5) equilateral, acute

 Total Problems: Total Correct: Score: 61

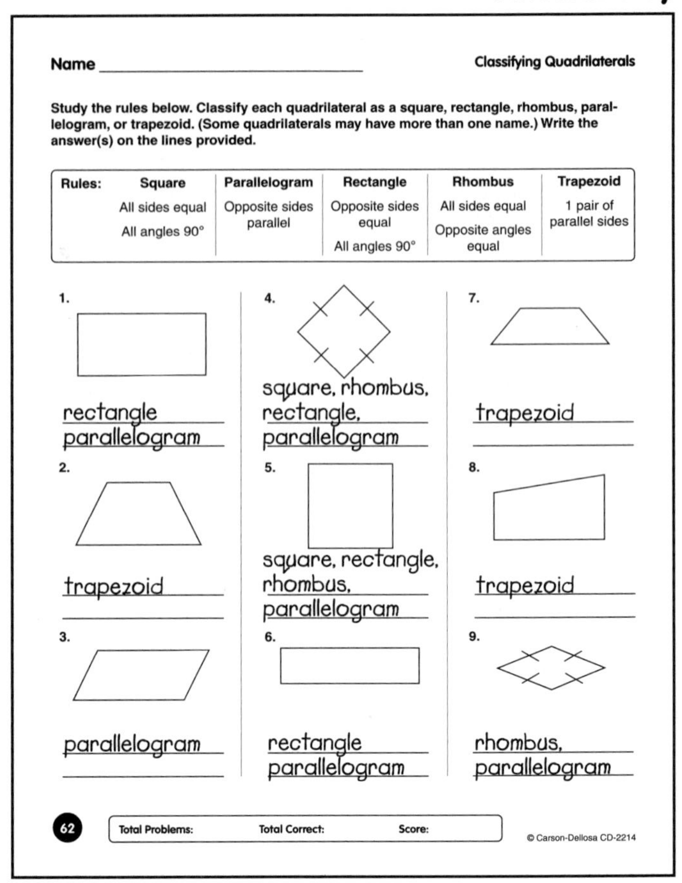

Name ____________________ **Classifying Quadrilaterals**

Study the rules below. Classify each quadrilateral as a square, rectangle, rhombus, parallelogram, or trapezoid. (Some quadrilaterals may have more than one name.) Write the answer(s) on the lines provided.

Rules:	Square	Parallelogram	Rectangle	Rhombus	Trapezoid
	All sides equal All angles 90°	Opposite sides parallel	Opposite sides equal All angles 90°	All sides equal Opposite angles equal	1 pair of parallel sides

1. rectangle, parallelogram
2. trapezoid
3. parallelogram
4. square, rhombus, rectangle, parallelogram
5. square, rectangle, rhombus, parallelogram
6. rectangle, parallelogram
7. trapezoid
8. trapezoid
9. rhombus, parallelogram

62 Total Problems: Total Correct: Score:

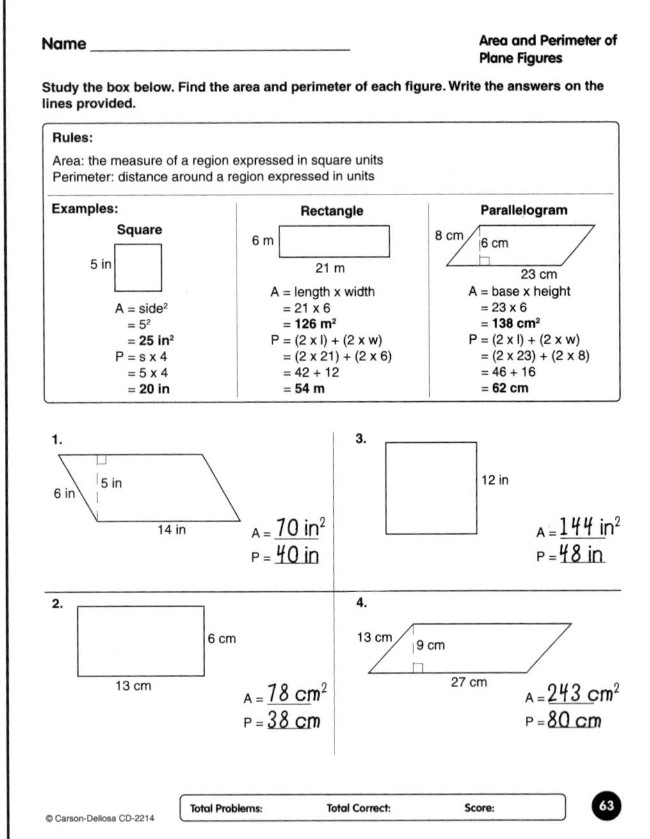

Name ____________________ **Area and Perimeter of Plane Figures**

Study the box below. Find the area and perimeter of each figure. Write the answers on the lines provided.

Rules:
Area: the measure of a region expressed in square units
Perimeter: distance around a region expressed in units

Examples:

Square	Rectangle	Parallelogram
5 in	6 m, 21 m	8 cm, 6 cm, 23 cm
A = side² = 5² = **25 in²**	A = length x width = 21 x 6 = **126 m²**	A = base x height = 23 x 6 = **138 cm²**
P = s x 4 = 5 x 4 = **20 in**	P = (2 x l) + (2 x w) = (2 x 21) + (2 x 6) = 42 + 12 = **54 m**	P = (2 x l) + (2 x w) = (2 x 23) + (2 x 8) = 46 + 16 = **62 cm**

1. (6 in, 5 in, 14 in) A = $70\ in^2$ P = 40 in
2. (6 cm, 13 cm) A = $78\ cm^2$ P = 38 cm
3. (12 in) A = $144\ in^2$ P = 48 in
4. (13 cm, 9 cm, 27 cm) A = $243\ cm^2$ P = 80 cm

 Total Problems: Total Correct: Score: 63

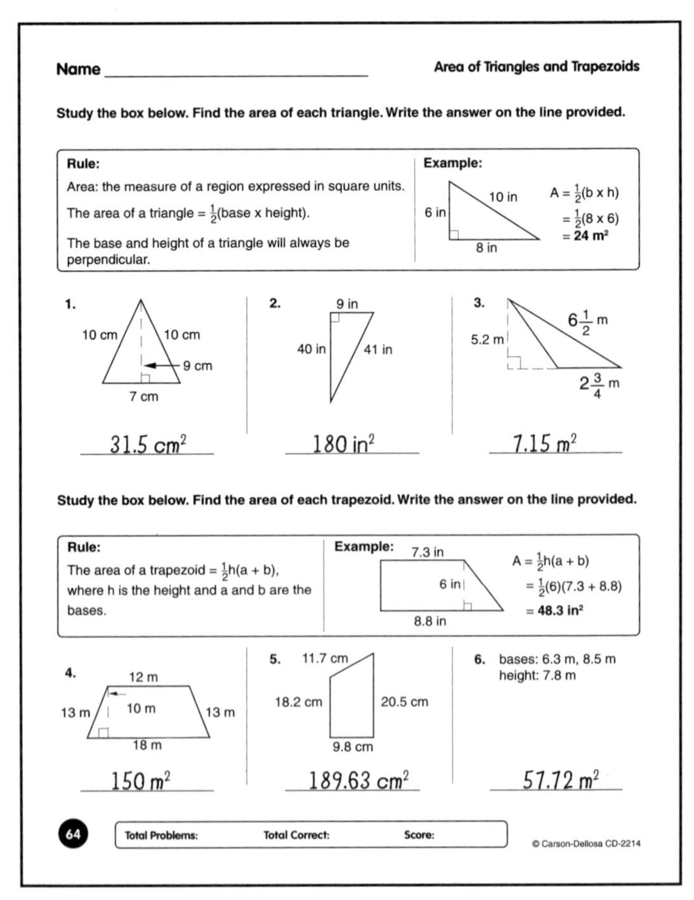

Name ____________________ **Area of Triangles and Trapezoids**

Study the box below. Find the area of each triangle. Write the answer on the line provided.

Rule:
Area: the measure of a region expressed in square units.
The area of a triangle = $\frac{1}{2}$(base x height).
The base and height of a triangle will always be perpendicular.

Example: (6 in, 10 in, 8 in) A = $\frac{1}{2}$(b x h) = $\frac{1}{2}$(8 x 6) = **$24\ m^2$**

1. (10 cm, 10 cm, 9 cm, 7 cm) $31.5\ cm^2$
2. (9 in, 40 in, 41 in) $180\ in^2$
3. ($6\frac{1}{2}$ m, 5.2 m, $2\frac{3}{4}$ m) $7.15\ m^2$

Study the box below. Find the area of each trapezoid. Write the answer on the line provided.

Rule:
The area of a trapezoid = $\frac{1}{2}$h(a + b), where h is the height and a and b are the bases.

Example: (7.3 in, 6 in, 8.8 in) A = $\frac{1}{2}$h(a + b) = $\frac{1}{2}$(6)(7.3 + 8.8) = **$48.3\ in^2$**

4. (12 m, 13 m, 10 m, 13 m, 18 m) $150\ m^2$
5. (11.7 cm, 18.2 cm, 20.5 cm, 9.8 cm) $189.63\ cm^2$
6. bases: 6.3 m, 8.5 m; height: 7.8 m — $57.72\ m^2$

64 Total Problems: Total Correct: Score:

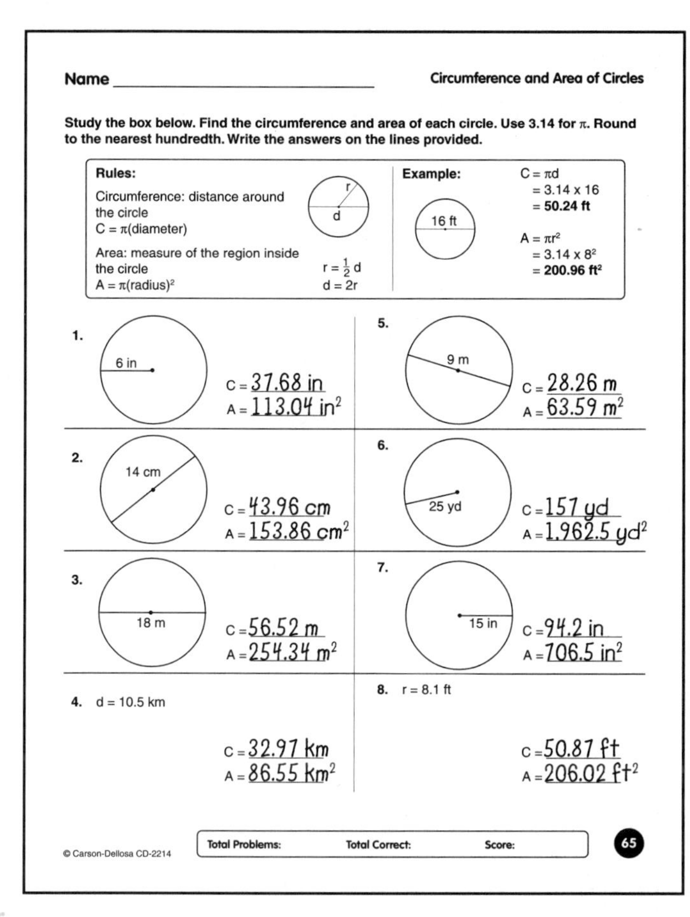

Name ______________________ **Circumference and Area of Circles**

Study the box below. Find the circumference and area of each circle. Use 3.14 for π. Round to the nearest hundredth. Write the answers on the lines provided.

Rules:
Circumference: distance around the circle
$C = \pi(\text{diameter})$
Area: measure of the region inside the circle
$A = \pi(\text{radius})^2$
$r = \frac{1}{2}d$
$d = 2r$

Example: (16 ft)
$C = \pi d$
$= 3.14 \times 16$
= 50.24 ft
$A = \pi r^2$
$= 3.14 \times 8^2$
= 200.96 ft²

1. 6 in — C = 37.68 in, A = 113.04 in²
2. 14 cm — C = 43.96 cm, A = 153.86 cm²
3. 18 m — C = 56.52 m, A = 254.34 m²
4. d = 10.5 km — C = 32.97 km, A = 86.55 km²
5. 9 m — C = 28.26 m, A = 63.59 m²
6. 25 yd — C = 157 yd, A = 1,962.5 yd²
7. 15 in — C = 94.2 in, A = 706.5 in²
8. r = 8.1 ft — C = 50.87 ft, A = 206.02 ft²

Total Problems: Total Correct: Score: 65

© Carson-Dellosa CD-2214

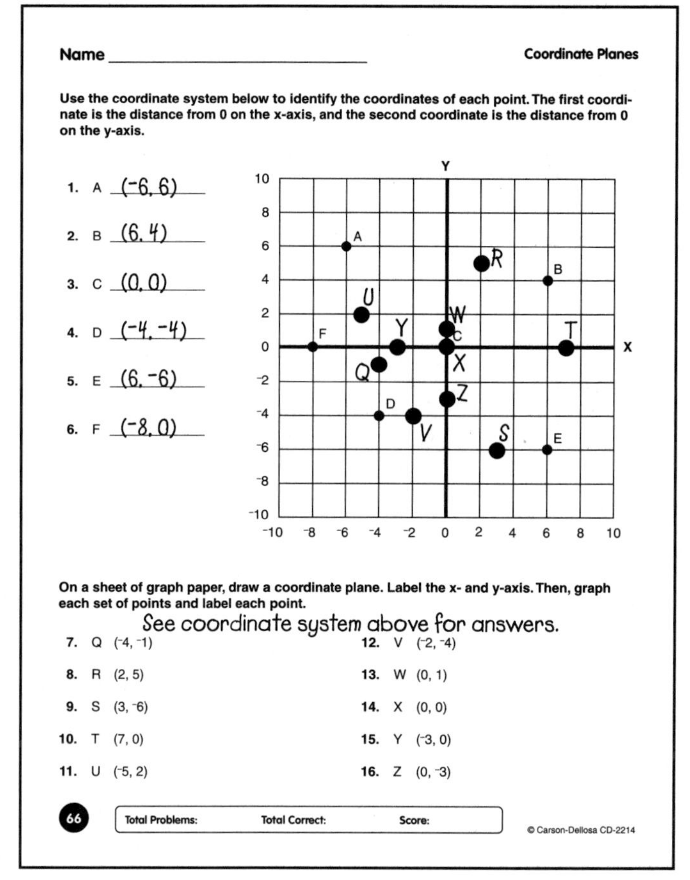

Name ______________________ **Coordinate Planes**

Use the coordinate system below to identify the coordinates of each point. The first coordinate is the distance from 0 on the x-axis, and the second coordinate is the distance from 0 on the y-axis.

1. A (−6, 6)
2. B (6, 4)
3. C (0, 0)
4. D (−4, −4)
5. E (6, −6)
6. F (−8, 0)

On a sheet of graph paper, draw a coordinate plane. Label the x- and y-axis. Then, graph each set of points and label each point.

See coordinate system above for answers.

7. Q (−4, −1)
8. R (2, 5)
9. S (3, −6)
10. T (7, 0)
11. U (−5, 2)
12. V (−2, −4)
13. W (0, 1)
14. X (0, 0)
15. Y (−3, 0)
16. Z (0, −3)

66 Total Problems: Total Correct: Score:

© Carson-Dellosa CD-2214

Name ______________________ **Problem Solving with Geometry**

Solve each word problem. Show your work and write the answer in the space provided.

1. To bisect an angle means to cut it in half. Write which type of angle is formed if each of the following angles are bisected:
   A. acute — acute
   B. right — acute
   C. obtuse — acute
   D. straight — right

2. Is it possible for a trapezoid to have 2 equal sides? Draw a figure to support your answer.
   Yes.
   Figures will vary.
   Example:

3. Which quadrilaterals have 4 congruent sides?
   square, rhombus

4. Find the area and circumference of a circle whose radius is 12 meters. What would happen to its area and circumference if you were to double its radius?
   $a = 3.14(12^2) = 452.16 \text{ m}^2$ | $c = 3.14(24) = 75.36 \text{ m}$
   area quadruples; circumference doubles

5. The area of a parallelogram is bh. The area of a triangle is $\frac{1}{2}$bh. Write 2 sentences describing their relationship. (bh = base x height)
   Since a triangle is half a parallelogram, its area will be half that of a parallelogram.

6. Is a square also a rectangle? Is it a parallelogram? Explain your answer.
   Yes. Yes.
   A square is a rectangle because opposite sides are equal and parallel, and there are four right angles. A square is a parallelogram because opposite sides are parallel and equal.

© Carson-Dellosa CD-2214 Total Problems: Total Correct: Score: 67

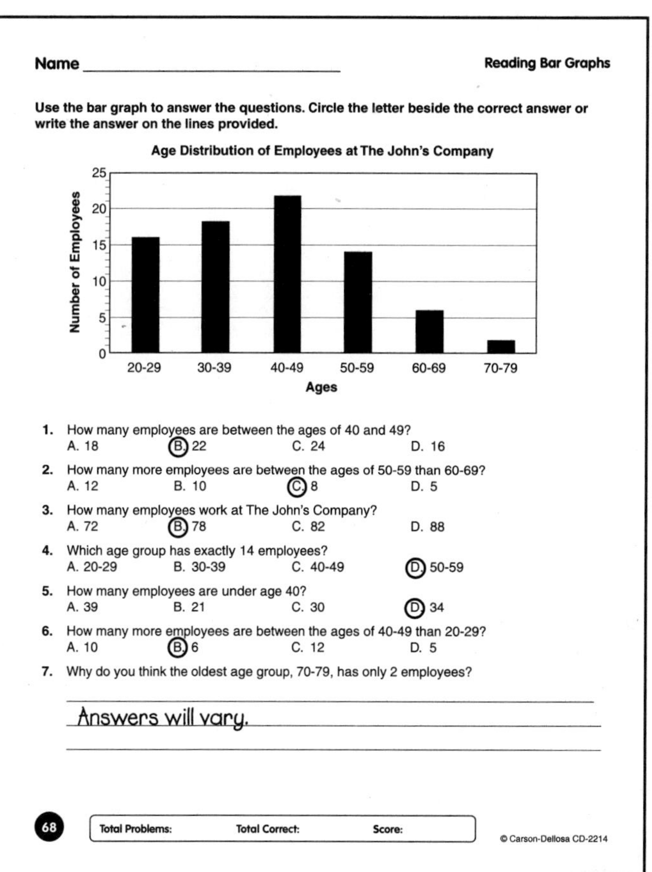

Name ______________________ **Reading Bar Graphs**

Use the bar graph to answer the questions. Circle the letter beside the correct answer or write the answer on the lines provided.

Age Distribution of Employees at The John's Company

1. How many employees are between the ages of 40 and 49?
   A. 18 **(B.) 22** C. 24 D. 16
2. How many more employees are between the ages of 50-59 than 60-69?
   A. 12 B. 10 **(C.) 8** D. 5
3. How many employees work at The John's Company?
   A. 72 **(B.) 78** C. 82 D. 88
4. Which age group has exactly 14 employees?
   A. 20-29 B. 30-39 C. 40-49 **(D.) 50-59**
5. How many employees are under age 40?
   A. 39 B. 21 C. 30 **(D.) 34**
6. How many more employees are between the ages of 40-49 than 20-29?
   A. 10 **(B.) 6** C. 12 D. 5
7. Why do you think the oldest age group, 70-79, has only 2 employees?
   Answers will vary.

68 Total Problems: Total Correct: Score:

© Carson-Dellosa CD-2214

Name ______________________ **Reading Line Graphs**

Use the line graph to answer the questions. Circle the letter beside the correct answer or write the answer on the lines provided.

Checked Out Books

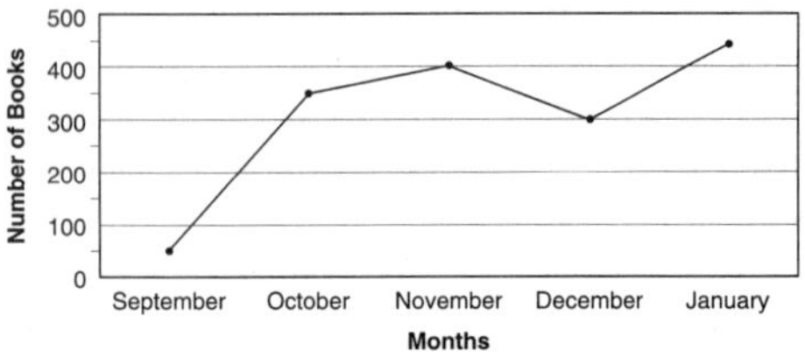

Number of Books: 0, 100, 200, 300, 400, 500

Months: September, October, November, December, January

1. How many books were checked out during the month of October?
   A. 300 (B) 350 C. 400 D. 450
2. How many more books were checked out in January than December?
   (A) 150 B. 175 C. 200 D. 225
3. During which month were 300 books checked out?
   A. October B. November (C) December D. January
4. Why do you think so few books were checked out in September?
   Answers will vary.
5. The librarian would like an average of 350 books checked out per month. Has her goal been reached so far? Support your answer.
   No. An average of 310 books per month have been checked out.
6. If there were 1,700 books in the library, what percentage of books were checked out in November?
   24% of the books

Total Problems: Total Correct: Score:

Name ______________________ **Reading Tables**

Use the tables to answer the questions. Write the answer on the line provided.

Mrs. Weber's Class

Number of Students	Number of Books Read
Marti	7
Terry	12
Rachel	9
Miquel	19
Kerren	8
Ernie	4
Stephanie	16

1. What is the greatest number of books a student has read?
   19
2. Who has read the fewest number of books?
   Ernie
3. What was the average number of books read by each student?
   10.7 books

Knights Basketball Team

Player	Points
Perez	9
Johnson	8
Duval	17
Blanchard	25
Schultz	16

4. Which player scored the most points?
   Blanchard
5. If the opposing team scored 61 points, did the Knights win or lose the game?
   The Knights won.

Color of Shoes Sold in One Hour

Color	Pairs Sold
Black	12
White	9
Red	3
Blue	6

6. How many red pairs of shoes were sold?
   3 pairs
7. How many more white, red, and blue shoes were sold than black?
   6 pairs
8. Which shoe color should the store order the most? Justify your answer.
   black because more were sold

Total Problems: Total Correct: Score:

Name ______________________ **Making Line Graphs**

Use the data in each table to make a line graph. Draw the graph in the space provided.

1. **Growth of Mindy's Baby**

Day	Weight (in Pounds)
July 1	12
August 1	14
September 1	17
October 1	18
November 1	20

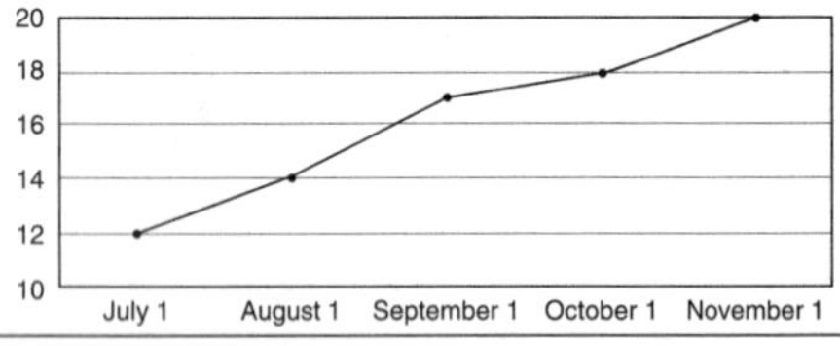

Graph axes: 10, 12, 14, 16, 18, 20; July 1, August 1, September 1, October 1, November 1

2. **Monthly Electric Bill**

Month	Amount
January	$108
February	$110
March	$95
April	$63
May	$48

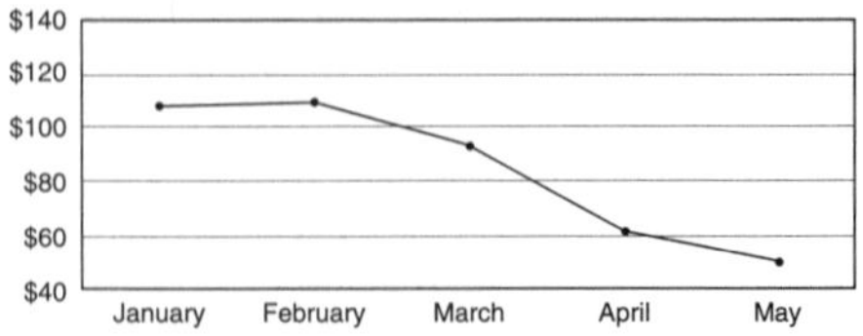

Graph axes: $40, $60, $80, $100, $120, $140; January, February, March, April, May

Total Problems: Total Correct: Score:

Name ______________________ **Stem and Leaf Plots**

Study the box below. In the space provided, make a stem and leaf plot for each set of data.

Rule: In a stem and leaf plot, the digits to the left of the line have the greater place value and are called stems. Digits to the right of the line represent digits in the ones place and are called leaves.

Example: 41, 42, 50, 39, 53, 48, 51, 38, 36, 48

stem	leaf
3	689
4	1288
5	013

1. 22, 21, 32, 42, 45, 55, 56, 24, 24, 26

stem	leaf
2	1 2 4 4 6
3	2
4	2 5
5	5 6

2. 89, 87, 99, 100, 95, 72, 78, 88, 88, 97, 98

stem	leaf
7	2 8
8	7 8 8 9
9	5 7 8 9
10	0

3. 210, 215, 217, 208, 229, 225, 216, 225, 205

stem	leaf
20	5 8
21	0 5 6 7
22	5 5 9

Use the following stem and leaf plot to answer the questions.

Height (in inches) of the basketball team

stem	leaf
6	168
7	001257
8	0

4. What is the height, in feet, of the shortest player? the tallest? 5 feet, 1 inch; 6 feet, 8 inches
5. Which height is the most frequent? 70 inches
6. How many people are on the team? 10
7. Where do most of the heights fall? in the 70s
8. Is there a player who is exactly 7 feet tall? no

Total Problems: Total Correct: Score:

Name ______________________ **Measures of Central Tendency**

Study the box below. Find the mean, median, and mode of each set of data. Round to the nearest tenth. Write the answers in the space provided.

Rules:	**Example:**
Mean (average): Add the numbers and divide by the total number in the set.	17, 15, 35, 10, 21, 11
Median (middle number): Place the numbers in order. Find the middle number. If there is not one middle number, average the two in the middle.	Mean: (17 + 15 + 35 + 10 + 21 + 11) ÷ 6 = 109; 109 ÷ 6 = 18.16; Rounds to **18.2**
Mode (most frequent): Find the number that occurs most frequently.	Median: The two middle numbers are 15 and 17. Reorder numbers: 10, 11, 15, 17, 21, 35; (15 + 17) ÷ 2 = **16**
	Mode: **No mode**

1. 30, 42, 50, 51, 40, 45, 50
   mean: 44
   median: 45
   mode: 50

2. 65, 65, 90, 60, 80, 62, 80, 62, 62
   mean: 69.6
   median: 65
   mode: 62

3. 34, 33, 39, 37, 29, 31, 36, 34
   mean: 34.1
   median: 34
   mode: 34

4. 235, 245, 330, 235, 320, 325, 435
   mean: 303.6
   median: 320
   mode: 235

Tell whether the mean, median, or mode would be the best measure for each given situation. Explain your answer.

5. Would you use mean, median, or mode to describe the typical selling price of a bicycle?
   median; Explanations will vary.

6. Would you use mean, median, or mode to determine the most popular toy sold at a store?
   mode; Explanations will vary.

Total Problems: Total Correct: Score:

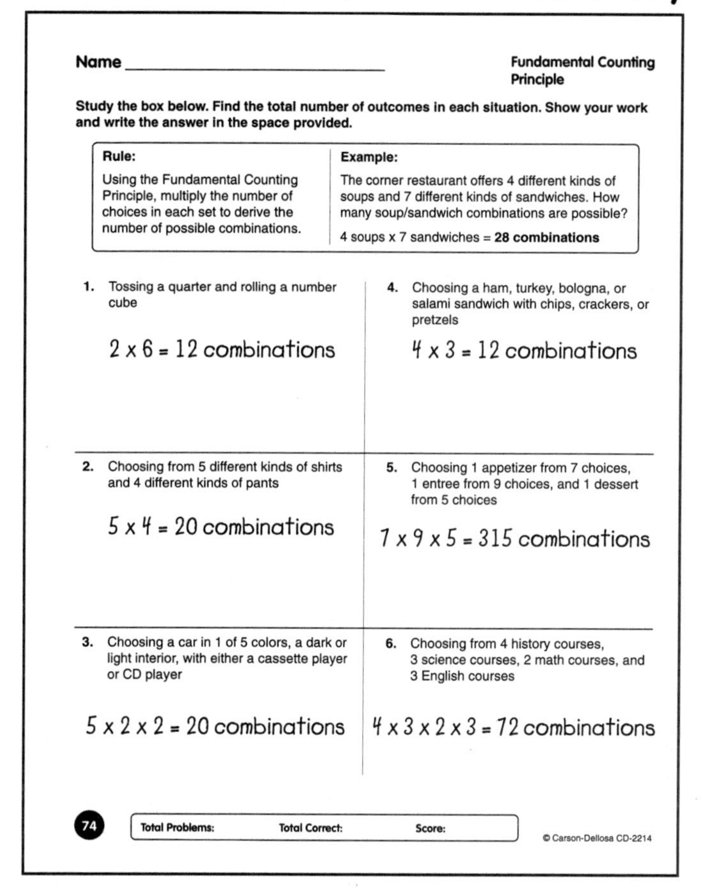

Name ______________________ **Fundamental Counting Principle**

Study the box below. Find the total number of outcomes in each situation. Show your work and write the answer in the space provided.

Rule:	**Example:**
Using the Fundamental Counting Principle, multiply the number of choices in each set to derive the number of possible combinations.	The corner restaurant offers 4 different kinds of soups and 7 different kinds of sandwiches. How many soup/sandwich combinations are possible? 4 soups x 7 sandwiches = **28 combinations**

1. Tossing a quarter and rolling a number cube
   2 x 6 = 12 combinations

2. Choosing from 5 different kinds of shirts and 4 different kinds of pants
   5 x 4 = 20 combinations

3. Choosing a car in 1 of 5 colors, a dark or light interior, with either a cassette player or CD player
   5 x 2 x 2 = 20 combinations

4. Choosing a ham, turkey, bologna, or salami sandwich with chips, crackers, or pretzels
   4 x 3 = 12 combinations

5. Choosing 1 appetizer from 7 choices, 1 entree from 9 choices, and 1 dessert from 5 choices
   7 x 9 x 5 = 315 combinations

6. Choosing from 4 history courses, 3 science courses, 2 math courses, and 3 English courses
   4 x 3 x 2 x 3 = 72 combinations

Total Problems: Total Correct: Score:

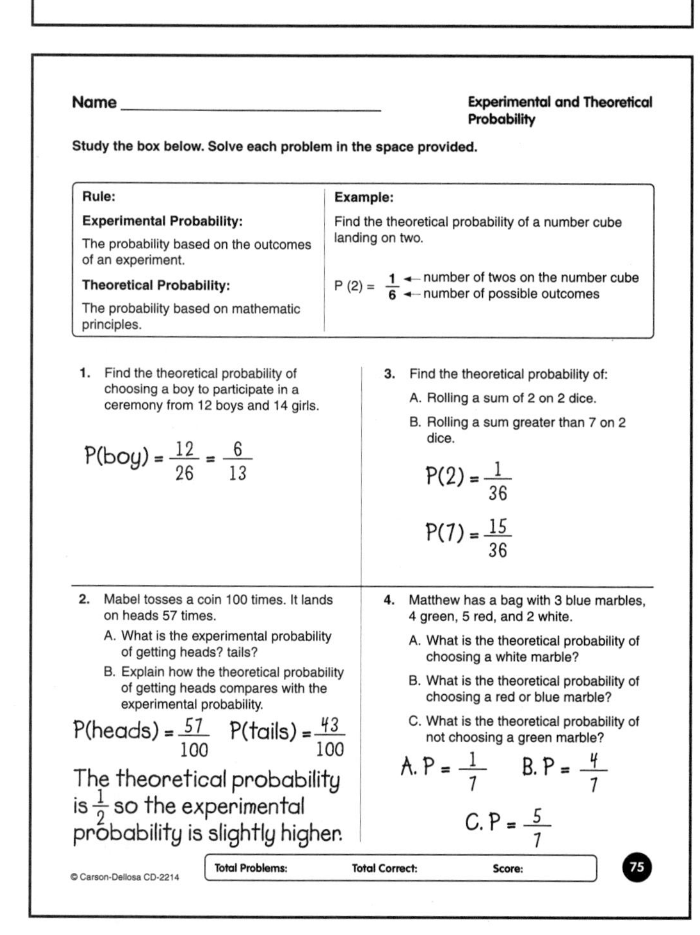

Name ______________________ **Experimental and Theoretical Probability**

Study the box below. Solve each problem in the space provided.

Rule:	**Example:**
Experimental Probability: The probability based on the outcomes of an experiment.	Find the theoretical probability of a number cube landing on two.
Theoretical Probability: The probability based on mathematic principles.	P (2) = $\frac{1}{6}$ ← number of twos on the number cube / ← number of possible outcomes

1. Find the theoretical probability of choosing a boy to participate in a ceremony from 12 boys and 14 girls.
   $P(boy) = \frac{12}{26} = \frac{6}{13}$

2. Mabel tosses a coin 100 times. It lands on heads 57 times.
   A. What is the experimental probability of getting heads? tails?
   B. Explain how the theoretical probability of getting heads compares with the experimental probability.
   $P(heads) = \frac{57}{100}$ $P(tails) = \frac{43}{100}$
   The theoretical probability is $\frac{1}{2}$ so the experimental probability is slightly higher.

3. Find the theoretical probability of:
   A. Rolling a sum of 2 on 2 dice.
   B. Rolling a sum greater than 7 on 2 dice.
   $P(2) = \frac{1}{36}$
   $P(7) = \frac{15}{36}$

4. Matthew has a bag with 3 blue marbles, 4 green, 5 red, and 2 white.
   A. What is the theoretical probability of choosing a white marble?
   B. What is the theoretical probability of choosing a red or blue marble?
   C. What is the theoretical probability of not choosing a green marble?
   A. $P = \frac{1}{7}$ B. $P = \frac{4}{7}$
   C. $P = \frac{5}{7}$

Total Problems: Total Correct: Score:

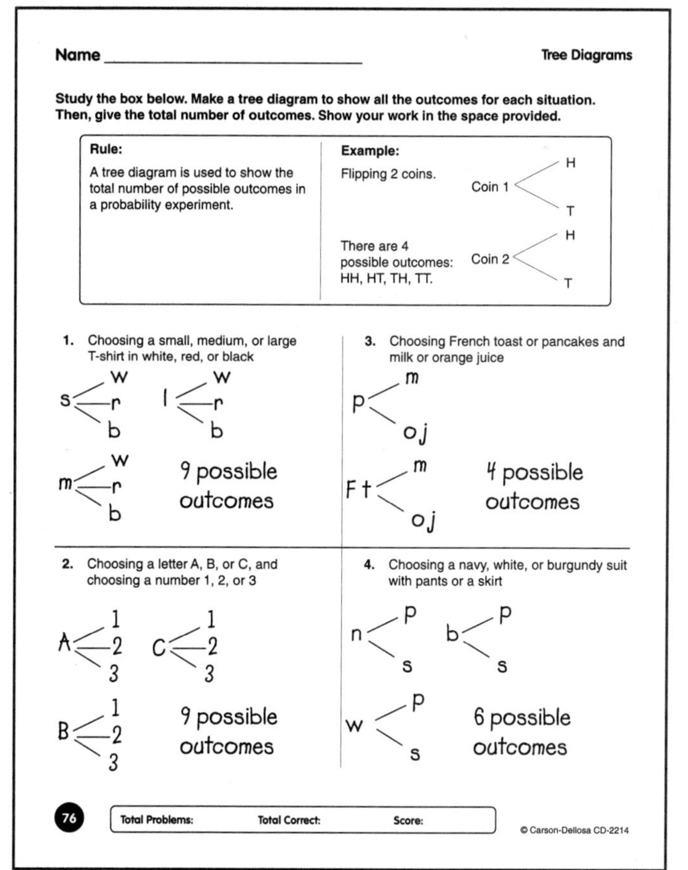

Name ______________________ **Tree Diagrams**

Study the box below. Make a tree diagram to show all the outcomes for each situation. Then, give the total number of outcomes. Show your work in the space provided.

Rule:	**Example:**
A tree diagram is used to show the total number of possible outcomes in a probability experiment.	Flipping 2 coins. Coin 1 < H, T; Coin 2 < H, T. There are 4 possible outcomes: HH, HT, TH, TT.

1. Choosing a small, medium, or large T-shirt in white, red, or black
   s < w, r, b; l < w, r, b; m < w, r, b
   9 possible outcomes

2. Choosing a letter A, B, or C, and choosing a number 1, 2, or 3
   A < 1, 2, 3; C < 1, 2, 3; B < 1, 2, 3
   9 possible outcomes

3. Choosing French toast or pancakes and milk or orange juice
   p < m, oj; Ft < m, oj
   4 possible outcomes

4. Choosing a navy, white, or burgundy suit with pants or a skirt
   n < p, s; b < p, s; w < p, s
   6 possible outcomes

Total Problems: Total Correct: Score:

Name ____________________ **Combinations with Probability**

Study the box below. Find the number of combinations. Write the answer in the space provided.

Rule:

A **combination** is an arrangement of items in which order does not matter.

A **permutation** is an arrangement of items in a particular order.

Example:

5 players from 9 — If you have 9 players and only 5 can play at a time, how many different combinations of 5 players are there?

$\frac{9 \cdot 8 \cdot 7 \cdot 6 \cdot 5}{5 \cdot 4 \cdot 3 \cdot 2 \cdot 1} = \frac{15{,}120}{120}$ ← Number of permutations of 9 players taken 5 at a time / ← Number of permutations of 5 players

$\frac{15{,}120}{120} = \mathbf{126}$ There are 126 combinations of players.

1. 4 movies from a list of 7

$\frac{7 \cdot 6 \cdot 5 \cdot 4}{4 \cdot 3 \cdot 2 \cdot 1} = 35$

2. 4 numbers from the numbers 1-5

$\frac{5 \cdot 4 \cdot 3 \cdot 2}{4 \cdot 3 \cdot 2 \cdot 1} = 5$

3. 3 numbers from 5, 10, 15, 20

$\frac{4 \cdot 3 \cdot 2}{3 \cdot 2 \cdot 1} = 4$

4. 2 letters from A, B, C, D, and E

$\frac{5 \cdot 4}{2 \cdot 1} = 10$

5. 4 girls from a group of 8

$\frac{8 \cdot 7 \cdot 6 \cdot 5}{4 \cdot 3 \cdot 2 \cdot 1} = 70$

6. 2 bows from 6 bows

$\frac{6 \cdot 5}{2 \cdot 1} = 15$

7. 5 shirts out of 12 shirts

$\frac{12 \cdot 11 \cdot 10 \cdot 9 \cdot 8}{5 \cdot 4 \cdot 3 \cdot 2 \cdot 1} = 792$

8. 3 applications from a choice of 15

$\frac{15 \cdot 14 \cdot 13}{3 \cdot 2 \cdot 1} = 455$

Name ____________________ **Problem Solving with Probability**

Solve each word problem. Show your work and write the answer in the space provided.

1. Suppose your teacher gives you a 4-question true/false quiz. How many sets of guesses are possible?

$4^2 = 16$ sets of guesses

2. Mark is choosing from among 4 brands of golf clubs, 2 kinds of golf balls, and 3 types of golf bags. In how many different ways can he buy a set of golf clubs, golf balls, and a golf bag?

$4 \times 2 \times 3 = 24$ different ways

3. A soft drink vending machine contains 2 soda buttons, 2 diet soda buttons, a ginger ale button, a root beer button, and an orange soda button. If you were to randomly choose a drink without looking, what would be the theoretical probability of choosing a diet soda?

$P(ds) = \frac{2}{7}$

4. In how many ways can a baseball coach choose 3 pitchers from a group of 5 able pitchers?

$\frac{5 \cdot 4 \cdot 3}{3 \cdot 2 \cdot 1} = 10$ different ways

5. During the first semester of Camille's literature class, each student must read and write about a short story, poem, and play. If there are 27 short stories, 18 poems, and 2 plays in her literature book, how many different selections can be made?

$27 \times 18 \times 2 = 972$ different ways

6. Steph decides to personalize her license plate. If the first 3 characters must be numbers between 1 and 9 (without repeating) and the next 3 characters must be a letter (any letter A-Z without repeating), how many possible ways can she create her license plate?

$\frac{9 \cdot 8 \cdot 7}{3 \cdot 2 \cdot 1} + \frac{26 \cdot 25 \cdot 24}{3 \cdot 2 \cdot 1} = 2{,}684$

2,684 possible ways